RANDOMNESS

RANDOMNESS

Deborah J. Bennett

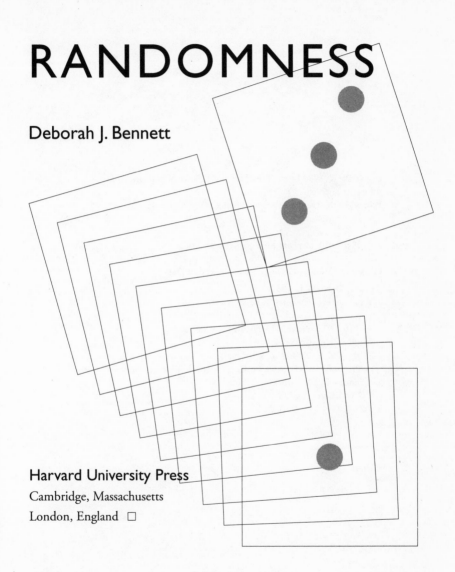

Harvard University Press

Cambridge, Massachusetts

London, England □

First Harvard University Press paperback edition, 1999

Library of Congress Cataloging-in-Publication Data

Bennett, Deborah J., 1950–
    Randomness / Deborah J. Bennett.
        p.   cm.
    Includes bibliographical references and index.
    ISBN 0-674-10745-4 (cloth)
    ISBN 0-674-10746-2 (pbk.)
    1. Probabilities—Popular works.   2. Probabilities—History.
3. Chance—Popular works.   I. Title.
QA273.15.B46   1998
519.2—DC21   97-35054

Acknowledgments

I greatly appreciate the assistance provided by the libraries and librarians at New York University Bobst Library, New York City Public Library, and Jersey City State College Forrest A. Irwin Library.

A Separately Budgeted Research grant from Jersey City State College provided me with much needed release time so that I could pursue this project. I would like to thank Kenneth Goldberg of New York University, George Cauthen of the Centers for Disease Control and Prevention, J. Laurie Snell, editor of *Chance News,* and Chris Wessman of Jersey City State College. Thanks also to Susan Wallace Boehmer for helping this book take shape, and special thanks to my brother, Clay Bennett, for the fabulous artwork in Figures 3, 4, 6, 9, 11, 17, and 18.

I am indebted to my editor, Michael Fisher, who has been a constant mentor and enthusiastic supporter during the writing process.

Finally I am particularly grateful to my husband, Michael Hirsch, without whom I could not have undertaken this effort. His patience, love, and understanding have sustained me.

Contents

RANDOMNESS

Chance Encounters

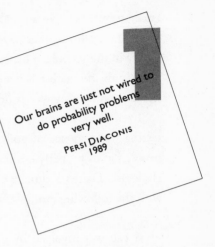

Our brains are just not wired to do probability problems very well.

PERSI DIACONIS
1989

Everyone has been touched in some way by the laws of chance. From shuffling cards for a game of bridge, to tossing a coin at the start of a football game, to awaiting the outcome of the Selective Service draft lottery, to weighing the risks and benefits of knee surgery, most of humanity encounters chance daily. The statistics that describe our probabilistic world are everywhere we turn: One-third don't survive their first heart attack. The chance of a DNA match is 1 in 100 billion. Four out of every 10 marriages in America end in divorce. Batting averages, political polls, and weather predictions are pervasive, but an understanding of the concepts underlying these statistics and probabilities is not.

Misconceptions abound, and certain concepts seem to be particularly problematic. To even the mathematically enlight-

2 ened, some issues in probability are not so intuitive. Despite curriculum reforms that have emphasized the teaching of probability in the schools, most experienced teachers would probably agree with the math teacher who commented, "Teaching statistics and probability well is not easy."[1]

Even in very serious decision-making situations, such as assessing the evidence of guilt or innocence during a trial, most people fail to properly evaluate objective probabilities. The psychologists Daniel Kahneman and Amos Tversky illustrated this with the following example from their research:

> A cab was involved in a hit and run accident at night. Two cab companies, the Green and the Blue, operate in the city. You are given the following data:
>
> (a) 85% of the cabs in the city are Green and 15% are Blue.
>
> (b) A witness identified the cab as Blue. The court tested the reliability of the witness under the same circumstances that existed on the night of the accident and concluded that the witness correctly identified each one of the two colors 80% of the time and failed 20% of the time.
>
> What is the probability that the cab involved in the accident was Blue rather than Green?[2]

A typical answer is around 80 percent. The correct answer is   **3**
around 41 percent. In fact, the hit-and-run cab is more likely to
be Green than Blue.

Kahneman and Tversky suspect that people err in the hit-
and-run problem because they see the base rate of cabs in the
city as incidental rather than as a contributing or causal factor.
As other experts have pointed out, people tend to ignore, or at
least fail to grasp, the importance of base-rate information because it "is remote, pallid, and abstract," while target information is "vivid, pressing, and concrete."[3] In evaluating the eyewitness's account, "jurors" seem to overrate the eyewitness's
likelihood of accurately reporting this specific hit-and-run event,
while underrating the more general base rate of cabs in the city,
because the latter information seems too nonspecific.

Base-rate misconceptions are not limited to the average person without an advanced mathematics education. Sophisticated
subjects have the same biases and make the same mistakes—
when they think intuitively. In a study at a prominent medical
school, physicians, residents, and fourth-year medical students
were asked the following question:

If a test to detect a disease whose prevalence is one in a
thousand has a false positive rate of 5 percent, what is the
chance that a person found to have a positive result actu-

4

ally has the disease, assuming you know nothing about the person's symptoms or signs?[4]

Almost half of the respondents answered 95 percent. Only 18 percent of the group got the correct answer: about 2 percent. Those answering incorrectly were once again failing to take into account the importance of the base-rate information, namely, (only) 1 person among 1000 tested will have the disease.

The commonsense way to think mathematically about the problem is this: Only 1 person in 1000 has this disease, as compared with about 50 in 1000 who will get a false positive result (5 percent of 999). It is far more likely that any one person who tests positive will be one of the 50 false positives than the 1 true positive. In fact, the odds are 1 in 51 that any one person who tests positive actually has the disease, and that translates into only a 2 percent chance, even in light of the positive test.

Another way to state the issue is that the chances of having this disease go from 1 in 1000 when one takes the test to 1 in 51 if a person gets a positive test result. That's a big jump in risk, to be sure, but it's a far cry from the 95 out of 100 chance many people erroneously believe they have after a positive test.

False positives are not human errors or lab errors. They happen because screening tests are designed to be overly sensitive in picking up people who deviate from some physiological norm, even though those people do not have the disease in question. In order to be sensitive enough to pick up most people who have tuberculosis, for example, skin tests for TB infection will always yield a positive result for around 8 percent of people who do not have the infection but who have other causes for reaction to the test; if 145 people are screened, roughly 20 will test positive. Yet only 9 of these 20 will turn out to have TB infection.[5]

The rate of false positives can be reduced by making screening tests less sensitive, but often this just increases the percentage of false negatives. A false negative is a test result that indicates no disease in a person who actually has the disease. Because false negatives are usually considered more undesirable than false positives (since people who get a false negative will not receive prompt treatment), the designers of screening tests settle on a compromise—opting for a very small percentage of false negatives and a somewhat larger percentage of false positives than we might prefer. In the case of tuberculosis, whereas roughly 7.5 percent of people tested will receive a false positive result, only 0.69 percent (roughly 1 person out of every 145 screened) will get a false negative test result. In other words, out of 145 people

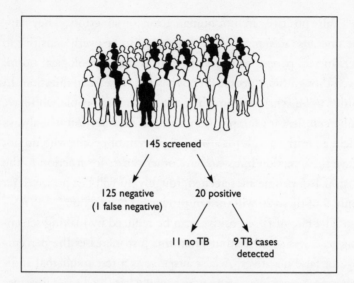

145 screened

125 negative   20 positive
(1 false negative)

11 no TB   9 TB cases
detected

FIGURE 1 Really well, or really sick? If a person has a routine screening test for tuberculosis, she or he has a 10 in 145 (about 7 percent) chance of having the infection at the time of the test. If the result comes back positive, the patient's odds of having TB go up to 9 in 20 (45 percent). If the result comes back negative, the patient still has a 1 in 125 chance of having the disease (about 0.8 percent); the original risk has been drastically reduced, but not eliminated, by the doctor's "good news."

screened for TB using this method, 9 cases of the disease will be
detected and 1 case will remain undetected (see Figure 1).

Considering that even highly educated medical personnel
can make errors in understanding probabilistic data of this kind,
we should not be at all surprised that probability often seems to
be at odds with the intuitive judgments of their patients and
other ordinary people.

In addition to base-rate misconceptions, psychologists have
shown that people are subject to other routine fallacies in evaluating probabilities, such as exaggerating the variability of chance
and overattending to the short run versus the long run.[6] For
example, the commonly held notion that, on a coin toss, a tail
should follow a string of heads is erroneous. Children seem
particularly susceptible to this fallacy. Jean Piaget and Barbel
Inhelder, who studied the development of mathematical thinking in children and whose work will be described frequently
in the following chapters, pointed out that "by contrast with
[logical and arithmetical] operations, chance is gradually discovered."[7]

One would think that the experiences acquired over a life-
time ought to solidify some correct intuitions about statistics
and probability. Intuitive ideas about chance do seem to precede
formal ideas, and, if correct, are an aid to learning; but if incorrect, they can hinder the grasp of probabilistic concepts. Kahne-

8 man and Tversky have concluded that statistical principles are not learned from everyday experience because individuals do not attend to the detail necessary to gain such knowledge.[8]

Not surprisingly, over the course of our species' history, acquiring an understanding of chance has been extremely gradual, paralleling the way an understanding of randomness and probability develops in an individual (if it does). Our human dealings with chance began in antiquity, as we will see in Chapters 2 and 3. Archaeologists have found dice, or dice-like bones, among the artifacts of many early civilizations. The practice of drawing lots is described in the writings of ancient religions, and priests and oracles foretold the future by "casting the bones" or noting whether an even or odd number of pebbles, nuts, or seeds was poured out during a ceremony. Chance mechanisms, or randomizers, used for divination (seeking divine direction), decision making, and games have been discovered throughout Mesopotamia, the Indus valley, Egypt, Greece, and the Roman Empire. Yet the beginnings of an understanding of probability did not appear until the mid-1500s, and the subject was not seriously discussed until almost one hundred years later. Historians have wondered why conceptual progress in this field was so slow, given that humans have encountered chance repeatedly from earliest times.

The key seems to be the difficulty of understanding random- **9**
ness. Probability is based on the concept of a random event, and
statistical inference is based on the distribution of random samples. Often we assume that the concept of randomness is obvious, but in fact, even today, the experts hold distinctly different
views of it.

This book will examine randomness and several other notions
that were critical to the historical development of probabilistic
thinking—and that also play an important role in any individual's understanding (or misunderstanding) of the laws of chance.
We will investigate a series of ideas over the course of the following chapters: ➤ Why, from ancient times to today, have people
resorted to chance in making decisions? ➤ Is a decision made
by random choice a fair decision? ➤ What role has gambling
played in our understanding of chance? ➤ Are extremely rare
events likely in the long run? ➤ Why do some societies and
individuals reject randomness? ➤ Does true randomness exist?
➤ What contribution have computers made to modern probabilistic thinking? ➤ Why do even the experts disagree about
the many meanings of randomness? ➤ Why is probability so
counter-intuitive?

We all have some notion about the "chances" of an event

occurring. We come to the subject of probability with some intuition about the topic. Yet, as the eminent eighteenth-century mathematician Abraham De Moivre pointed out long ago, problems having to do with chance generally appear simple and amenable to solution with natural good sense, only to be proven otherwise.[9]

Why Resort to Chance?

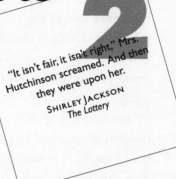

Everyday randomizers are not very sophisticated. To settle a dispute over which child gets to ride in the front seat of the car, for example, a parent may resort to a game called "drawing straws." One child holds two thin sticks or broom straws in her hand, with the ends concealed, and the other child chooses. The child with the shorter straw wins. Many adult entertainments—from old-fashioned cake walks and Friday Night Bingo to school raffles and million-dollar jackpots—are driven by simple lotteries. Spinners turn up on children's board games, among teenagers playing spin-the-bottle, and in Las Vegas gambling casinos. Dice, which are among the oldest randomizers known, are still popular today among a range of ages and ethnic groups.

Although hand games, lots, spinners, dice, coins, and cards are not very complicated devices, our attitudes about using them are a great deal more complex. When primitive societies needed to make a selection of some sort, they often resorted to randomizers for three basic reasons: to ensure fairness, to prevent dissension, and to acquire divine direction. Modern ideas about the use of chance in decision-making also invoke issues of fairness, dispute resolution, and even supernatural intervention, though we usually think of these concepts somewhat differently today.

Interestingly, these three reasons are exactly the ones given by children when asked by psychologists why they used counting-out games, such as one potato/two potato. Ninety percent of the time children responded that counting out gave them an equal chance of being selected. Other reasons given were to avoid friction and to allow some kind of magical or supernatural intervention.[1] Clearly, the idea of fairness is an important intuitive element in children's notions of randomness. Of course children eventually learn that counting-out games are not really fair: the choosing can be manipulated by speeding up or slowing down the verse, or by changing the starting point for counting out. Once they figure this out, children generally move on to better methods of randomization.

When chance determines the outcome, no amount of intelligence, skill, strength, knowledge, or experience can give one player an advantage, and "luck" emerges as an equalizing force. Chance is a fair way to determine moves in some games and in certain real-life situations; the random element allows each participant to believe, "I have an opportunity equal to that of my opponent."

Some of these situations properly fall under the heading of "problem resolution and problem conclusion." If we are ambivalent about the alternatives and truly don't care which decision is made, but acknowledge that some decision must be made, the simple toss of a coin may well relieve us of the responsibility, time, and thought required to analyze the alternatives.

Or perhaps we do care which decision is made but find ourselves at an impasse—as in an argument over whether a particular play is legal or illegal in a backyard game of touch football. If both sides were to hold their positions firmly, play could not resume. Although the decision made by tossing a coin may not be fair in the sense of determining who was right and who was wrong, any decision that both parties have an equal chance of receiving may be better than no decision at all. So a second important reason we resort to chance is to avoid dissension and get on with the show.

Because children have such a strong sense of fairness, their exploration of the concept of chance often leads to the question, "Is chance fair?" And in some situations, as children quickly discover, chance is not fair. There are many occasions when "taking turns" would seem to be the fair way to resolve a conflict

14   or make a choice, and selection by chance would seem quite unfair.

For example, the random selection of a particular child for some daily or weekly honor or privilege may impress young children as fair at first, but it is possible that one or more individuals might be selected more than once, while others are never selected. In fact, any outcome is possible with random selection, as we will see. What has already happened does not affect the chances that a particular outcome will happen the next time; four heads in a row do not make a tail more likely on the next throw. To older children who can understand this, random selection may not seem fair at all—not fair in the same sense that taking turns is fair. To those not selected, it's not much consolation to know that if the "long run" is long enough, they will be selected equally often.

When blame or punishment or an onerous duty is to be assigned, choices made by chance may seem absolutely unjust. In an essay read on public radio several years ago about the unfairness of chance, a writer related an experience she had as a young child in Catholic school. Lent, the Christian season of atonement, is often spent by denying oneself some luxury or vice while pondering all the acts one should be repentant for. At the writer's particular Catholic school, a lottery was being held to determine which luxury each child would be obliged to give up for the 40 days of Lent.

Hesitantly, each student drew a slip of paper and announced the vice that she must forgo for 40 days: ice cream, candy bars, comic books, and the like. As the writer mixed the slips, made her random draw, and read the selection she had made, a sharp intake of breath was heard and then a hush filled the room—TELEVISION. In a lottery where all the prizes are bad but one is *much worse than the others,* random assignment may not seem fair at all.

The Shirley Jackson short story, "The Lottery," had a profound effect on me when I first read it as a young woman. The story takes place on the day of the annual lottery in a New England town. Drama builds as the entire citizenry anticipates the drawing in a tradition that has been abandoned by the surrounding towns as archaic. Gradually the reader gets the feeling that the townspeople don't want their lot to be selected. The lot of Mrs. Hutchinson is finally drawn: she has been selected for the town's ritual stoning.[2]

An interesting reference to the injustice of a one-time decision made by lottery can be found in the Talmud. The rarely questioned Hebrew notion that everything happens at the direction of the Lord sees an exception in the story of Achan. In order to determine the party guilty of a particular offense, Joshua cast lots, and the guilty lot fell on Achan, who said: "Joshua, dost thou convict me by a mere lot? Thou and Eleazar the Priest are the two greatest men of the generation, yet were I to cast lots upon you, the lot might fall on one of you." Achan objected to his guilt being determined by lottery, complaining that, guilty or innocent, the lot must of necessity fall on *someone.*

But Achan was alone in his concern about the use of a random mechanism for such an important determination. To all others, it was obvious that the lot manifested God's judgment and not blind chance. Divine intervention had correctly identified the guilty party. Achan eventually confessed his guilt and was chastised to cease his irreverent questioning of the use of lots. The leaders who decided the guilty by lottery had ample psychological ammunition on their side to make the suspected party uncomfortable enough to confess a sin—perhaps uncomfortable enough to confess a sin he did not commit.[3]

Important decisions, we moderns usually think, should be judicious and rely on logic rather than chance. When the outcome of the decision is of little consequence, or we find ourselves in a situation where we simply cannot choose between alternatives, then and only then are most people willing to leave the decision to chance.

It has not always been this way, however. In ancient times,

randomizers that eliminated any human element of logic or skill
played a major role in games and in important life decisions.

Ancient Randomizers in Games of Chance

Though not always recognized or acknowledged as such, chance
mechanisms have been used since antiquity: to divide property,
delegate civic responsibilities or privileges, settle disputes among
neighbors, choose which strategy to follow in the course of
battle, and drive the play in games of chance.

Several gaming boards originating from ancient Babylonia
have been uncovered as a result of archaeological excavations
in Ur. One, dating circa 2700 B.C., was found complete with
playing pieces or "men." It is believed that the game was driven
by some sort of chance mechanism, though none was found. At
the site of the palace at Knossos in Crete, an intricate inlaid
gaming board, dating to the Minoan civilization of 2400–2100
B.C., was discovered. Although no playing pieces or dice were
unearthed, the game is believed to be similar to backgammon.[4]

Boards found at later Babylonian, Assyrian, Palestinian, and
other sites resemble a game which originated in ancient Egypt
and has much in common with modern cribbage. Pegs that fit
into holes mark the path around the board. A board game
similar to Snakes and Ladders or a primitive form of backgammon and dating to circa 1878–1786 B.C. was excavated from an

Egyptian tomb in Thebes, along with playing pieces in the form of ivory hounds and jackals.[5]

Why have so many ancient boards turned up without the accompanying dice? Perhaps other implements of greater antiquity were used, and those objects are simply not recognizable to us today as being dice. If shells or pebbles or other objects found in nature were thrown as dice, archaeologists may not have made the connection. Or perhaps the objects were made of materials such as wood or plant matter that have long since decomposed. Another possibility is that some personal devices such as fingers were used to determine number and drive the game. In such instances, no artifacts identifiable as randomizers would be found.

The earliest six-sided dice known have come from the East. Made of baked clay and dating circa 2750 B.C., one die was found in excavations of ancient Mesopotamia in Northern Iraq, Tepe Gawra. Dice dating from about the same period and made from the same material were found in all strata at the Mohenjo-Daro excavation in the Indus valley—including rectangular prism stick dice and triangular stick dice, as well as cubical dice (see Figure 2). At both sites, the faces of the six-sided dice were marked with pips (dots). In those times pips were used probably because there was as yet no system of nu-

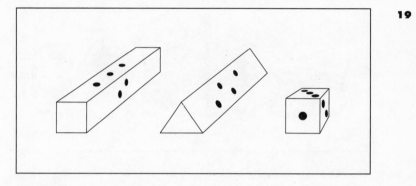

FIGURE 2 Ancient rectangular prism stick dice, triangular stick dice, and cubical dice.

merical notation, but the tradition of using pips rather than numbers has been retained in most modern dice.[6]

Six-sided dice have been found in Egypt from excavations dating to 1320 B.C., but dice-like bones used as chance devices have turned up in much earlier Egyptian sites. These particular bones, found also in the later Greek and Roman civilizations, are from the heel of a hoofed animal such as a deer, calf, sheep, or goat and are called *tali* (Latin) or *astragali* (Greek). The astragalus has four distinctly different long flat faces, the only ones on which it will rest when tossed, and two small rounded

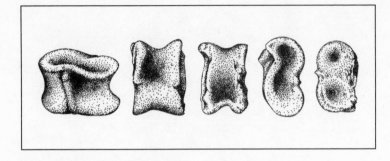

FIGURE 3 The astragalus and its four different faces.

ends. Of the four flat faces, two are narrow and flat and two are broad, with one broad side slightly convex and the other slightly concave. Since each side of the astragalus looked different (see Figure 3), it was not necessary to mark the sides. If they were marked with pips, the sides were scored 1, 3, 4, and 6.

Greek and Roman games of chance were played with four astragali. The lowest throw, known as "the dogs," resulted when all ones were thrown. In the highest throw, called the Venus-throw, four different sides came up. The astragalus bone was imitated in ivory, bronze, glass, and agate and was used in religious ceremonies as well as in games. The Greek word for die or cube is *kubos,* from which is derived the Talmudic word for dice, *kubya,* and the English word *cube.* The Arabic name for the astragalus, as well as the cubical die, is *kab*, meaning ankle. Some experts have taken all this to mean that the astragalus was the ancestor of the cubical die, but any direct connection still eludes us.[7]

Game boards, playing pieces, and astragali have been found in a number of excavations in ancient Egypt—from a burial chamber in Thebes circa 2040–1482 B.C., to the tomb of Queen Hatasu, circa 1600 B.C. In the tomb of Tutankhamen (born 1358 B.C.) archaeologists found a gaming board, together with playing pieces, ivory astragali, and two-sided stick dice painted black on one side and white on the other.[8]

The earliest chance devices from Egypt may have been much simpler than even the four-sided astragalus. Illustrated in a Third Dynasty (circa 2800 B.C.) tomb, and displayed side by side with a gaming board, are staves or reeds having a convex and a concave side, which may have been tossed, like heads or tails, to determine moves on the board. Ivory staves (dated circa 3000 B.C.), marked on one side and plain on the other, along with playing pieces and rods apparently used for scoring or counting, have been found in excavations near Thebes (see Figure 4). The Metropolitan Museum of Art in New York City has on display such staves or throwing sticks from Thebes dating circa 1420 B.C. Accompanying a gaming board and playing pieces are nine throwing sticks, stained red on one side, which

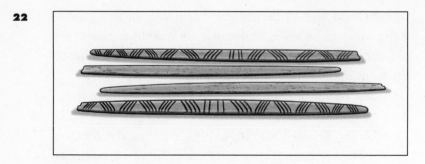

FIGURE 4 Two-sided throwing sticks or staves like those found in excavations of Thebes, 3000 B.C.

were apparently thrown to determine the moves of the pieces. Some archaeologists also think that certain beads, inscribed on one side, may have served as two-sided dice.[9]

Play at the games of odd-or-even and *morra* is depicted on the walls of the ancient Egyptian tombs of Beni Hassan, circa 2000 B.C. (see Figure 5). Morra—a popular hand game still played today in Italy—was called *micare digitis* by the ancient Romans. Two people simultaneously extend some, none, or all of the fingers of the right hand while calling out a guess as to the sum. In *The Lives of the Caesars,* Suetonius describes an incident when the cruel Emperor Augustus required a father and son to cast lots or play morra to decide whose life should be spared.[10]

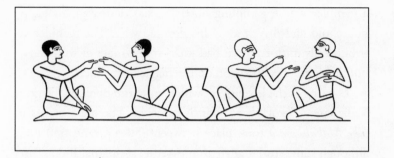

FIGURE 5 Play at *morra* and odd-and-even illustrated on the walls of the tomb of Beni Hassan, 2000 B.C., as documented by the Greek historian, Herodotus, circa 450 B.C.

Emperor Augustus was devoted to gaming and even provided his dinner guests with a sum of money "in case they wished to play at dice [tali] or at odd-and-even during the dinner." Odd-or-even was a favorite game among the Greeks and Romans, who called it *par impar ludere.* One person would hold a number of beans, nuts, coins, or astragali in his hand, and the stake was won if the opponent correctly guessed whether the number of items was even or odd.[11]

One of the earliest written documents about chance mechanisms in gaming is from among the Vedic poems of the *Rgveda Samhita.*[12] Written in Sanskrit circa 1000 B.C., this poem or song, called the "Lament of the Gambler," is a monologue by a gambler whose gambling obsession has destroyed his happy household and driven away his devoted wife. The gambler attributes his misfortune to bad luck—not to fate or to a god—but he blames his obsession on the magic of the dice.

Gaming is the activity on which an entire book of the great Vedic epic, *The Mahabharata,* is centered. The main events in *The Mahabharata* took place between 850 B.C. and 650 B.C., although some stories relate to pre-Vedic, aboriginal India. Parts of the epic were written as early as 400 B.C., and others were written as late as A.D. 400.[13] From *The Mahabharata,* we know the dice were rolled or thrown onto a board, and that the dice were made from brown nuts, probably from the vibhitaka tree. These hard nuts are almost round but have five slightly flattened sides; they are about the size of hazel nuts or nutmeg. It appears that the game was based on a simple call of odd or even and had nothing to do with the fact that the dice had five sides. A large quantity of dice (nuts) were strewn over the dicing carpet. They were not marked, or if they were, it was irrelevant to the rules of the game. The two competing dicers agreed upon the number of games to be played and the stakes to be wagered. The first player declared odd or even and grasped a handful of dice. They were subsequently thrown down and counted. If the player guessed right, he won the stakes and the game was over. If not, he lost his bet, and the other person took his turn and became the player.[14]

In one segment of the epic (the story of Nala) there is evidence that skill at the vibhitaka game had something to do with the ability to count up a great number of nuts very quickly. In this tale, a man named Nala becomes the employee of a king, who claims, "Know that I know the secret of the dice and am expert at counting."[15] To demonstrate his talent at counting, the king counts the number of nuts on two large branches of a vibhitaka tree. His method of "counting" is first to estimate the ratio of leaves and nuts on the ground to those on the tree, and then to estimate the ratio of leaves to nuts. Finally, by estimating the number of leaves on the two branches, the king is able to say that there are 2195 nuts on the two branches. The nuts are counted, and of course he is exactly correct.

In theory this instantaneous counting would produce an expert at the even/odd vibhitaka game, but some scholars believe that the story of Nala was an idealization. Such skill was not really possible; it was a game of chance.[16] Others, however, have pointed out that there exist in modern India professional appraisers, called *kaniya,* who estimate the crop yield for landowners with uncanny accuracy. Because their reliability has been established from year to year, their results are rarely questioned.

Whether such estimation skills were possible or not, considering the antiquity of *The Mahabharata,* I am inclined to agree with Ian Hacking when he says that the story of Nala provides a "curious insight into the connection between dicing and sampling."[17]

Another ancient Indian game was played with wooden or ivory stick dice, called *pasakas*—rectangular prisms with four long flat scoring sides marked with pips. In fact, there were numerous types of dice known in ancient India—two-sided wooden chips, coins, or shells, five-sided nuts, four-sided pasakas, and six-sided cubical dice. The different types of dice may have existed concurrently, and it is not certain which came first. One scholar has speculated that the pips on the pasakas are a symbolic representation of the concrete vibhitaka nuts, and as such illustrate the evolution of dice from more primitive types of randomizers.[18]

We find a wide assortment of chance devices used in games among the American Indians. The games differed mainly in the choice of the dice themselves, as tribes were limited to available materials. With few exceptions, the dice had two faces—one concave and the other convex, or one face marked or painted in some distinguishing way. Two-sided dice were made of bone, wood, seeds, woodchuck or beaver teeth, walnut shells, hickory staves, crows' claws, or plum stones. The Eskimos used six-sided ivory and wooden dice, which were shaped like chairs, though only three sides counted. Among the Papago Indians, bison astragali were used as dice, with only two sides counting.[19]

The widespread distribution of such games in the New World seems to support the theory that these dice were very old, predating Columbus, and were not imported.[20] The guessing game of odd-or-even, often played with sticks, also appears to be aboriginal among American Indians.

From the earliest days of civilization, people have invented simple mechanisms for the purpose of removing human will, skill, and intelligence from their play and from serious decision making. Yet, paradoxically, as we will see in Chapter 3, generally the ancients believed that the outcome of events was ultimately controlled by a deity, not by chance. The use of chance mechanisms to solicit divine direction is called divination, and the steps taken to ensure randomness were intended merely to eliminate the possibility of human interference, so that the will of the deity could be discerned.

When the Gods Played Dice

In antiquity, whether chance mechanisms were being used for serious decision making or for games of chance, a strong belief existed that the gods controlled the outcome. The purpose of randomizers such as lots or dice was to eliminate the possibility of human manipulation and thereby to give the gods a clear channel through which to express their divine will. Even today, some people see a chance outcome as fate or destiny, that which was "meant to be."

Methods to secure the randomness of a lottery are mentioned as far back as Homer (circa 850 B.C.). In the *Iliad,* in which he relates the adventures of the Greeks during the Trojan War (supposedly in the twelfth century B.C.), a lottery is used to reveal who should cast the first spear in a duel between Menelaus and Paris. Paris' seduction of Menelaus' wife, Helen, had precipitated the Greeks' assault on Troy, but at the time of the duel the city had not yet fallen to the Greeks. The lots of the two men were put into a helmet (the lots had presumably been marked or designated in some way), the helmet was shaken, the people in attendance prayed to the gods, and the warrior whose lot was first shaken out by Hector—commander of the Trojan forces—was the one chosen to cast his spear first.

Though Homer makes it clear throughout the *Iliad* that the gods have absolute power to manipulate events, we find that the Greeks had set procedures to impose fairness in the lottery. In the line, "Hector of the shining helmet shook the lots, looking backward, and at once Paris' lot was outshaken," we see two interesting ideas.[1] First, the lots must be *shaken*—it appears that even with divine intervention, measures must be taken to prevent human interference. And second, Hector is *looking backward*, just as one might close one's eyes to be impartial in the selection.

In a later passage of the *Iliad*, a lottery was used to choose which Greek soldier would fight Hector himself. Each soldier "marked a lot as his own one lot," the lots were put into a helmet, the gods were prayed to, the lots were shaken, "and a lot leapt from the helmet."[2] Only the one who had marked it recognized the lot as his—not surprising, since the Greek alphabet came into use after the Trojan War.

It is not unusual in ancient literature to see references to the chance object's having animate qualities—leaping from the helmet in this case. Piaget and Inhelder, in their work on children's notions about chance, recount similar responses in the way children express their thoughts. When asked what the pattern of colors would look like when colored balls were mixed randomly, the youngest children remarked, "No one can be sure because we aren't balls," and "They know where to go because they're the ones who have to do it."[3]

Writing many centuries after Homer, around A.D. 98, the Roman historian Tacitus described the Teutonic method of divination by lots:

> For omens and the casting of lots they have the highest regard. Their procedure in casting lots is always the same. They cut off a branch of a nut-bearing tree and slice it into strips; these they mark with different signs and throw them completely at random onto a white cloth. Then the priest of the state, if the consultation is a public one, or the father of the family if it is private, offers a prayer to the gods, and looking up at the sky picks up three strips, one at a time, and reads their meaning from the signs previously scored on them.[4]

The reference to looking up at the sky prior to selecting the lots could have several interpretations. The priest may be seek-

ing to assure the onlookers that he was not interfering with
chance, just as Hector did when looking backward. Or, to the
contrary, as a magician uses misdirection to divert the audience's
attention away from sleight of hand, the diviner could be divert-
ing attention away from his interference. Or, in conjunction
with the prayer, looking skyward could provide an extra appeal
to the gods for guidance.

In addition to shaking the lots and looking backward, sometimes a child was used to select a lot. Cicero relates that the lots
at Praeneste in the temple of Fortuna were taken out only at the
direction of the goddess and then shuffled and drawn by a child.
The purity of children made them appropriate instruments of
divine will, since they were supposedly impervious to inten-
tional bias.[5]

Hebrew commentaries from the Middle Ages suggest still
other measures that could be taken to create randomness in a
lottery. Concerning the lot selection for the sacrificial scapegoat
on the Day of Atonement, Moses Maimonides writes,

Concerning the two lots: on one of them was written
"for the Lord," and on the other was written "for Azazel."
They might be made of any material: wood, stone or
metal. However, one was not to be large and the other
small, or one of silver and the other gold. Rather, both

were to be alike . . . Both lots were placed in a vessel that could contain two hands, so that one might put in both his hands without reaching purposely [for a particular lot] . . . The High Priest shook the urn and brought up in his two hands two lots for the two he-goats.

In Hebrew tradition, it was considered a good omen if the lot "for the Lord" came up in the priest's right hand.[6]

In the Old Testament, lots were drawn to select the scape-goat for atonement, to select a specific date for the sacrifice, to delegate authority, to assign responsibility, to select the residents of Jerusalem, and to identify the party guilty of some offense. When a single individual had to be selected from a large group, this was often accomplished through several stages of selections—what we might today call cluster sampling. First, a tribe was selected from among all tribes, then from the tribe a family was selected, and from the family an individual was selected. In these cases the lots represented individuals, families, or tribes.[7]

The most frequent use of lots in the Old Testament is to divide an inheritance of property or conquered lands or the spoils of war. After the area to be divided was surveyed and partitioned, the lots themselves probably were marked to represent particular parcels of land.[8] In this way the words denoting the "lot" came to have other meanings, such as a parcel of land, an assigned function, or one's destiny. The *Assyrian Dictionary* defines the word *isqu* to mean a "lot (as a device to determine a selection)" but also as a "share (a portion of land, income, property, or booty . . . assigned by lot)," and also as "fortune, fate, destiny."[9]

Talmudic references indicate that in some cases *two* urns of lots were used. In one were the lots of the participants and in the other were lots representing the boundary descriptions of the lands. One lot taken from each urn determined the allocation.[10] This may be a better plan for randomization than the one-urn set-up. From the results of the 1970 United States Selective Service draft lottery, it was determined that using selection of birthdays from one drum did not allow sufficient mixing to guarantee randomness. Some birthdays were more likely to get picked first and therefore get a low number (and probably be inducted), and other birthdays were more likely to get picked last and receive a high draft number (and probably avoid induction).

This happened because, before the drawing, birth dates had been added to the drum one month at a time in sequence, January first and December last. The insufficient mixing of the lots meant that birth dates near the end of the year were more likely to be picked first and therefore receive a lower draft number. When this bias in mixing was discovered, a new plan evolved which was based on drawings from two drums, one containing 365 birthdays and the other containing 365 draft numbers. One lot drawn from each drum determined the assignment of draft number to birth date.[11]

In A.D. 73 the Jewish defenders of Masada, rather than die at the hands of their enemy once the battle became hopeless, drew lots to determine who would carry out the mass suicide. Shards inscribed with the names of men have been found and are believed to be those used. Ten men were chosen by lot to become the executioners. After the executions took place, one was chosen by lot to kill the other nine executioners, after which he committed suicide.[12] Could a similar lottery have been used in the March 1997 mass suicide of the Heaven's Gate cult in southern California? Most likely we will never know.

The Old Testament indicates two reasons why a random mechanism, such as lots, might have been used to make life-and-death decisions. According to Proverbs 16:33, "The lot is cast into the lap; but the whole disposing thereof is of the Lord," that is, divine will directed the fall of the lot. This was a notion rarely questioned. Another more practical reason that decisions were made by lot is brought out by King Solomon in Proverbs 18:18, "The lot causeth contention to cease, and parteth between the mighty." In Biblical times the resort to chance was an agreed-upon way of making many decisions because it ended

dissension among opposing, often powerful, parties. Quarrel-
some rabbis, for example, frequently used lots to allocate daily
duties in the temple.[13]

In Mesopotamia, the casting of lots was used to make selections, in the belief that the deity directly affected, indeed manipulated, the lots. In ancient Assyria, the eponym of the year
was determined by lot. The king gave his own name to the first
year of his reign, and each subsequent year was named for an
official of the kingdom who was chosen by lot. One such lot, a
clay die, has been found for the year 833 B.C., and from it we
can see a direct enthusiastic appeal to the gods to affect the
outcome of the selection. The die is inscribed: "O great lord,
Assur! O great lord, Adad! this is the lot of Jahali, the chief
intendant of Shalmaneser, king of Assyria, [governor of] the city
of Kipsuni, of the countries . . . , the harbor director; make
prosper the harvest of Assyria and let it be bountiful in the
eponym [established] by his lot! May his lot come up!"[14]

Participants in early European lotteries, like the peoples of
ancient Mesopotamia, often invoked God and his saints for
assistance. A lottery craze seems to have overtaken the populace
beginning in the sixteenth century, and lotteries continued to
flourish during the seventeenth and eighteenth centuries. Typical of that time, a child (perhaps blindfolded) would draw a
numbered slip of paper from an urn or a wheel of fortune, or

perhaps a ball from a rotating hopper. The lottery was seen by rich and poor alike as an equalizing force, since the prize money was equally available to all who played.[15]

Chance and the Book of Changes

The *I Ching*, or *Book of Changes*, which is widely consulted to this day by the Chinese, began as an oracle involving instruments of chance. According to Chinese tradition, the *I Ching* was one of five books written or compiled by Confucius in the fifth or sixth century B.C.[16] It consists of 64 hexagrams (figures with six lines) and their associated interpretations. Through the use of some chance device, one eventually arrives at two of these hexagrams to predict one's particular fortune. Two hexagrams are necessary, because the book of "changes" relates to a change in moving from one hexagram to another.

Each hexagram is composed of 6 lines, wherein one of two symbols can occur (see Figure 6). Originally these two symbols were known as the light and the dark, but later became known as the *yin* and the *yang*.[17] Since each of six lines can occur in one of two ways (either yin or yang), a total of 64 different hexagrams is possible: 2 in the first line, 2×2 in the first two lines, $2 \times 2 \times 2$ in the first three lines, and so on until you reach $2 \times 2 \times 2 \times 2 \times 2 \times 2 = 64$.

I Ching specialists believe that the hexagram evolved by put-

yin   yang

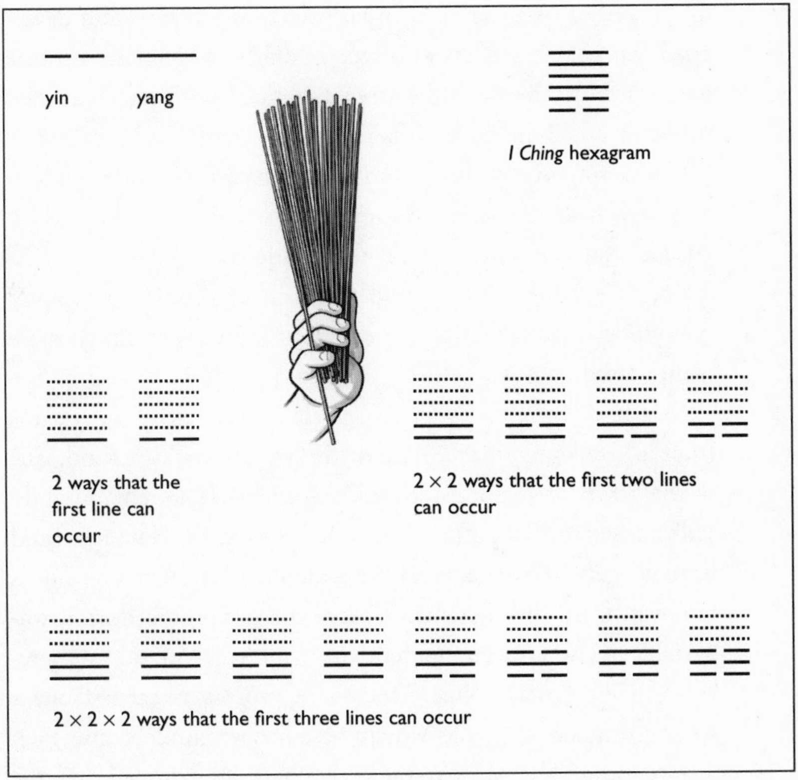

I Ching hexagram

2 ways that the
first line can
occur

2 × 2 ways that the first two lines
can occur

2 × 2 × 2 ways that the first three lines can occur

FIGURE 6 Consulting the *I Ching*.

ting together two trigrams. In all probability, the system developed from a primitive yes-no type of oracle, which later became more sophisticated with the arrangement of first three, and then six, such possibilities. It is therefore possible to arrive at a particular hexagram by using some type of two-outcome random mechanism six times. For example, the toss of a coin—heads for yin and tails for yang—could determine one line of the hexagram. Five more tosses would result in a complete hexagram symbol. In actual practice, the techniques used were much more complicated.[18]

Methods for consulting the oracle involve either 50 yarrow (milfoil) stalks or three coins. In the yarrow-stalk method, one of the 50 stalks is put aside and never used. The remaining 49 stalks are first divided into two piles at random. The left-hand heap is counted by fours and the remainder is noted; after one is removed from the right-hand heap, the same procedure is undertaken. The two remainders are summed, and this sum receives a numerical value. The entire process is carried out a second time on the stalks which were not remainders, and then a third time. The three numerical values are summed and the final sum results in either the yin or the yang that fills one line of the hexagram. This procedure is repeated six times to arrive at a completed six-line hexagram. Although this method of summing remainders is highly involved, it turns out that the chances of a yin or a yang occurring are equal.[19]

The Chinese method of casting an oracle using coins is much simpler than yarrow stalks, though never as simple as even/odd or heads/tails. Three ancient bronze Chinese coins, each with a hole in the middle and inscribed on one side, are thrown down at once. According to whether the inscribed side is facing up, a particular numerical value is assigned to the coin, and the numerical values are summed. The sum results in one line of the hexagram, and the procedure is repeated six times.[20] We may marvel at the elaborate complexity of the procedures undertaken to arrive at a simple yin or yang. Perhaps the process was long and drawn out in order to impress the participants of its legitimacy and seriousness.

The oracle or fortune that was arrived at through the *I Ching* was believed to be based on a partnership between man and god. One reference to divination by milfoil, written by the Chinese poet Su Hsun in the eleventh century A.D., reveals this belief:

And he took the milfoil. But in order to get an odd or even bunch in milfoil stalks, the person himself has to divide the entire bunch of stalks in two . . . Then we count the stalks by fours and comprehend that we count by fours; the remainder we take between our fingers and know that what is left is either one or two or three or four, and that we selected them. This is from man. But dividing all the stalks in two parts, we do not know [earlier] how many stalks are in each of them. This is from heaven.[21]

By his manipulation of the stalks, man unconsciously becomes an active participant in the oracle. This philosophy, at least by the eleventh century A.D., when Confucianism was enjoying a revival among Chinese intellectuals, suggests that man shaped the divination by his participation, while the random component ("We do not know how many stalks are in each of them") is divinely determined.

An interesting passage in another Confucian treatise, the *Shu Ching* (*Book of History* or *Book of Documents,* sixth century B.C.), suggests that maybe one should use one's own judgment in deciding whether to follow an oracle—a feature we have not seen elsewhere. The text explains how one can seek guidance through divination using milfoil stalks, or an even older method, using tortoise shells: "There are altogether seven kinds of divination, five of which are with a tortoise shell, and two with milfoil stems. This is to allow for doubts. And appoint these people and

let them predict (with the tortoise) and divine (with the milfoil).
If the three divine, follow the answer of at least the two who are
in agreement."[22]

These instructions are interesting because they admit the
possibility of error, or chance, or at least the likelihood that
different answers might occur. They certainly encourage the
seeking of a second (and third) opinion. The methods have
none of the divine infallibility of a single drawn lot that appears
in many other cultures. In fact, the petitioner is told not to rely
on one diviner or system of divination but rather to follow the
majority decision. Even after the diviners have arrived at their
predictions, the treatise continues to allow for doubt and to
encourage judgment: "But if you have great doubt, then deliberate upon it, turning to your own heart; deliberate, turning to
your retinue; deliberate, turning to the common people; deliberate, turning to the people who predict and divine."

This feature of a moral obligation to deliberate on the results
of the oracle, rather than to accept it at face value, distinguishes
the *I Ching* from other forms of divination; it is precisely this
feature of moral obligation which gradually changed the document from a divinatory to a philosophical text. In the introduction to his translation of the *I Ching*, Richard Wilhelm says that
although the *I Ching* may have originally been the type of oracle

which foretold one's fate, the first time a man refused to let the matter of his fate rest but asked, "What can I do to change it?" the book became a book of wisdom.[23]

Using chance to arrive at a particular passage in the *I Ching* is a form of rhapsodomancy: the seeking of guidance through the chance selection of a passage in a literary work. Another early form of rhapsodomancy is represented by the sibylline books. They were put into their present form in the sixth century A.D., although the first books were probably written around the sixth century B.C., when Greek oracles were at the height of their popularity. According to legend, there were a number of sibyls, the earliest of whom was from Persia; another was Jewish, reportedly Noah's daughter, from either Babylon or Egypt. The sibylline books were oracle verses, supposedly uttered by the sibyls in a prophetic frenzy.[24]

The books were held in great reverence and stored in the temple of Jupiter in Rome until it burned down in 83 B.C. In 76 B.C. a new collection of oracular verses was compiled, but those too later burned, in A.D. 405. This second collection was certainly written in Greek hexameter, and Cicero, writing circa 44 B.C., said that some of the verses were in the form of acrostics. Other ancients have suggested that the original verses were written in hieroglyphs and also mentioned the acrostic code.[25]

What the books were made of and how they were employed **43** may forever remain a mystery; it might be that the books were unrolled at random and a passage taken, or perhaps the passage was subject to the choice of the interpreters. These oracles, like some others, may have been written on loose leaves (or thin sheets of wood) which could be shuffled and a passage drawn at random. The fact that the oracles were vague and obscure, apt to apply to any number of situations, lends credence to this theory.

Not everyone felt comfortable with such a random, hap-hazard method of obtaining an oracle, however. Virgil's Helenus warns Aeneas of the capriciousness in the sibyl's method:

> You will find an ecstatic, a seeress, who in her ante
>   communicates
> Destiny, committing to leaves the mystic messages.
> Whatever runes that virgin has written upon the leaves
> She files away in her cave, arranged in the right order.
> There they remain untouched, just as she put them
>   away:
> But suppose that the hinge turns and a light draught
>   blows through the door,
> Stirs the frail leaves and shuffles them, never thereafter
>   cares she

> To catch them, as they flutter about the cave, to restore
> Their positions or reassemble the runes: so men who
>   come to
> Consult the sibyl depart no wiser, hating the place.[26]

Throughout the Middle East, Europe, and Asia, ancient civilizations turned to chance devices to reveal the will of the gods, and they took elaborate precautions to ensure that human participants could not influence the outcome. But just because one's fate was in the hands of the gods, that did not mean that fairness or justice would result. This cynical side of the human psyche was represented in the Roman goddess of fate, Fortuna. A personification of fickleness, she showed no signs of fairness, neither rewarding virtue nor punishing vice. Fortuna enjoyed a resurgence in popularity in Europe during the Middle Ages and the Renaissance, where she appeared in the works of Boethius, Dante, Boccaccio, Chaucer, and Machiavelli, among others.[27]

Literary and artistic references to Fortuna at times depict her as blind or blindfolded, showing a complete disregard for the virtuous, the rich, or the powerful. The English poet William Blake, in a note on his illustrations to Dante, wrote, "The Goddess Fortune is the devil's servant, ready to kiss any one's arse." She is described as both impartial and unfair—sometimes depicted as having many hands, taking away as easily as she **45** gives. Or, her right and left hands may represent good and evil fortune, respectively. At times, she has wings, as fortune is fleeting. She has been depicted as standing on a ball or a globe, or turning a wheel. In Roman tradition, Fortuna played dice games with men's lives, their very fate determined by the game's outcome.[28]

Figuring the Odds

Many adults and older children can grasp the fact that heads are as likely an outcome as tails in any given coin toss. They understand that the probability of winning the toss is 1 out of 2, no matter which way they call it. Similarly, they know that the chances of throwing a particular number with one die is 1 out of 6, and that the odds do not change whether that number is 1, 2, 3, 4, 5, or 6.

Yet when asked to calculate the probability of getting a particular sum when *two* dice are thrown, many adults would be at a loss. And estimating the chances of filling a full house in poker, or drawing an inside straight, is beyond the capacity of most people. It is not so easy to figure out the probability of a particular outcome when all possible outcomes are not equally probable.

The simplest type of random event is one whose outcomes

are equally likely—there is simply no way of knowing ahead of time which outcome will occur. The concept "equally likely" was undoubtedly difficult to grasp in early times, in large part because dice and coins usually were not symmetrically constructed. Certainly the ancients must have known that the sides of the astragalus were not equiprobable, since two sides were narrow and flat, and of the two broad sides, one was convex and the other concave. But the ability to formulate an idea of how likely the sides were would have been impossible for the ancients.

In modern times, it has been demonstrated that the probabilities are roughly 10 percent for each of the two narrow, flat sides of the astragalus and 40 percent for each of the two broad sides. But as Florence Nightingale David, who conducted the experiments behind these calculations, cautions, the probabilities "would undoubtably be affected by the kind of animal bone used, by the amount of sinew left to harden with the bone, [and] by the wear of the bone."[1]

By the early 1600s, Galileo had a clear idea of what we would today call a fair die, and he understood the concept of equally likely. As he described it, a fair die "has six faces, and when thrown it can equally well fall on any one of these."[2] To get a feeling for the difference between equal and unequal probabilities, let's first consider an experiment with equally likely outcomes—the roll of a fair six-sided die. Because the die is constructed symmetrically and balanced to ensure uniformity, each of the numbers is equally likely to face up when the die is rolled. On a single roll of a die, the number of pips showing can be 1, 2, 3, 4, 5, or 6. To put it formally, we would say that the *equally likely sample space*—the list of all possible outcomes—consists of the numbers 1 through 6.

If you tried to guess which number would face up when the die is rolled, you would have 1 chance in 6 of being right (see Figure 7, top). The probability of your being right is expressed as $\frac{1}{6}$. Since the probability of any particular side facing up is uniformly the same over all the numbers 1 through 6, the distribution of probabilities for a fair die is said to be *uniform*. This is also referred to as an equally likely sample space.

Now with a paintbrush and paint let's alter one of the faces of the fair die to illustrate an experiment with outcomes that are not equally likely. Suppose the side with one pip receives an additional pip, so that now when the die is rolled the number of pips facing up can be any of the five numbers 2 through 6. Though the die is still a fair die, so that each of the six sides is equally likely to face up when rolled, the chances of getting any particular result are not equally likely, and the distribution of probabilities is not uniform. While the probabilities of rolling a 3, 4, 5, or 6 are all still $\frac{1}{6}$, the probability of rolling a 2 is $\frac{2}{6}$, or

$\frac{1}{3}$, and the probability of rolling a 1 is zero (see Figure 7, middle).

In trying to explain nonuniform probabilities to his eighteenth-century readers, Abraham De Moivre, who penned the first modern book on probability in 1756, introduced a game in which each of the five players has 1 chance out of 5 of winning, or a probability of $\frac{1}{5}$.[3] The distribution of probabilities is uniform. De Moivre then says to imagine that from among the five persons who began the game with equal probabilities of winning, two must leave the game. If these two give their chances over to one of the remaining players, that player now has 3 chances in 5 of winning the sum, or a probability of $\frac{3}{5}$.

In experiments such as this one, an *event* is one or more outcomes in the sample space. A *simple event* is exactly one outcome in the sample space. A certain side facing up on the roll of a single die with six equally likely outcomes is an example of a simple event. A certain side facing up on the roll of the painted die with five (not equally likely) outcomes is also an example of a simple event. A certain *sum* facing up on the roll of two or more dice, however, is an example of a *compound event,* and calculating probabilities of such events is much more complex.

A compound event is an event involving two or more simple events. Compare the simple event of noting the total number of pips facing up on the roll of one die with the compound event

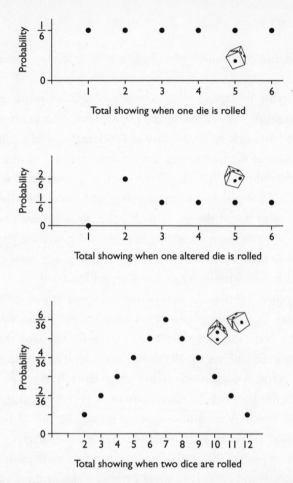

FIGURE 7 Probabilities of the total showing when one die, an altered die, and two dice are rolled. The roll of one die (top) is a simple event, and the probabilities are uniform. The roll of two dice (bottom) is a compound event, and the probabilities are nonuniform. The roll of one altered die (middle) is a simple event, but the probabilities are nonuniform.

of noting the total number of pips facing up on the roll of two dice. On the roll of one fair six-sided die, the number of pips showing can be 1 through 6, and each throw is equally likely. On the roll of two fair dice, the number of pips showing can be 2 through 12, but these sums are not equally likely (see Figure 7, bottom).

To understand why, let's imagine using colored dice, one red and one green. For each of the 6 possible throws on the red die, 6 are possible on the green die, for a total of $6 \times 6 = 36$ equally possible throws. But many of those throws yield the same sum. To make things even more complicated, different throws can result in the same two numbers. For example, a sum of 3 can occur when the red die shows 1 and the green die shows 2, or when the red die shows 2 and the green die shows 1. Thus the probability of throwing a total of 3 is 2 out of 36 possibilities, or $\frac{2}{36}$. A sum of 7, on the other hand, can be thrown 6 different ways—when red is 1 and green is 6; red is 6 and green is 1; red is 2 and green is 5; red is 5 and green is 2; red is 3 and green is 4; red is 4 and green is 3 (see Figure 8, top). Therefore the probability of throwing a 7 is $\frac{6}{36}$.

The fact that the sums on the faces of two dice are not equally likely is obscured by the compound nature of the event—one die *and* another die. This may be easier to visualize if one imagines rolling one die twice (rather than two dice at

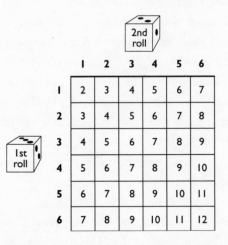

FIGURE 8 Possible sums on one roll of two dice (one green and one red; top), compared with the possible sums on two rolls of one die (bottom).

once) and then summing the results. The possible sums on two
rolls of one die are exactly the same as those on one roll of two
dice (see Figure 8, bottom). Therefore, the probability distribution for the total on two dice rolled at once is identical to that
for one die rolled twice. Frequently, a compound event can be
modeled by a sequence of simple events. One die rolled *N* times
may prove easier to visualize than *N* dice rolled at once.

To take another example, imagine that the room is dark and
you can't see a thing. In your drawer are two loose socks, identical except for color—one is red and one is blue. If you reach into
the drawer and withdraw one sock, are you more likely to get a
red or a blue sock? In this simple example, we can conclude that
the probability of getting a red and the probability of getting a
blue are equal. The outcomes of choosing a red sock and choosing a blue sock are equally likely.

Now let's complicate things by putting three socks in the
drawer—two are blue and one is red, identical except for color.
If you withdraw two socks, are you more likely to get two blues
(a match) or one blue and one red (no match) or are the outcomes equally likely? In fact, the chances of getting one of each
color are twice as likely as the chance of getting two blues. But
this fact is not at all obvious. You can get no match in two
ways—the red can be paired with either one of the two blues.

But you can get a match in only one way—the blues can only be paired with each other.

Imagine the whole process in slow motion. When you put your hand into the drawer to choose two socks, one of the three was touched (selected) first and then the other (see Figure 9). Suppose one of the blues was selected first—this is twice as likely to occur as first selecting a red. Once the blue is selected, we have an equal chance of getting a match or no match since both a red and a blue are left. Now suppose a red sock was selected first. At this point, you are guaranteed a no-match outcome, since there are only blues left to be selected with the red.

The socks-matching problem is an example of the importance of *sequence* in random selections. Even though the original question was posed as a problem involving the removal of two socks at once, the event is easier to analyze as the removal of two socks in sequence—first one and then the other. In the evolution of the theory of probability, the inability to recognize the sequential nature of random outcomes has been a major stumbling block, and it continues to plague learners even today.

Set versus Sequence

The objective of many two-dice games is to achieve a certain sum on the roll of two dice. But many early European dice

Chances 2 out of 3

Blue selected first.
Red and blue left.

Chance of match 50%
Chance of no match 50%

Chances 1 out of 3

Red selected first.
Two blues left.

Chance of match 0%
Chance of no match 100%

FIGURE 9 If two blue socks and one red sock are in a drawer and two are withdrawn, which is more likely, a match or one of each color?

56 games were played with three dice.[4] When three dice are thrown and the numbers of pips on the dice are summed, any one of the totals 3 through 18 can occur. Those sums can come about, however, in 216 (6 × 6 × 6) different ways. In other words, the equally likely sample space consists of 216 outcomes.

It appears that for centuries, learned men believed that there were only 56, rather than 216, possible three-dice outcomes. What they failed to recognize was the difference between a collection or *set* and a *sequence*. As a result of this misconception, they counted sets, when they should have been counting sequences.

The set {1,2,3} is identical to the set {3,2,1}; the order of the elements is immaterial. On the other hand, it is the elements *and* the order that determine the sequence (1,2,3). Indeed, this sequence is not the same as the sequence (3,2,1), or any other different ordering of the numbers 1, 2, and 3. When throwing three dice together, all one observes is the final *set* of numbers— the particular sequence which produced that set is hidden. If only the final outcome is of interest, then how it came about seems unimportant. *In computing probabilities, however, it is essential to know the number of equally likely ways an outcome can come about.*

Suppose that after tossing three dice we observe a throw of 1, 1, 2—in other words, the set outcome {1,1,2}. In fact there are 3 ways that outcome could have occurred—by rolling a $(1,1,2)$ sequence, a $(1,2,1)$ sequence, or a $(2,1,1)$ sequence. By contrast, the throw of 1, 2, 3, or set outcome {1,2,3}, can come about by rolling any one of 6 equally likely sequences: $(1,2,3)$, $(1,3,2)$, $(2,1,3)$, $(2,3,1)$, $(3,1,2)$, or $(3,2,1)$. The probability of observing a final outcome of two 1s and a 2 is 3 chances out of 216, or $\frac{3}{216}$. The probability of observing a final outcome of one 1, one 2, and one 3 is 6 chances out of 216, or $\frac{6}{216}$. If the three dice are rolled all at once, the idea that a final outcome may have occurred in a number of ways remains obscured. That is, the *sequential nature* of this random event is difficult to see, but seeing it is essential to calculating probabilities accurately.

Games employing three dice had been popular since the days of the Roman Empire; yet there is little evidence that a thorough understanding of the importance of sequence existed. The first known correct enumeration of equiprobable tosses of three dice is attributed to Richard de Fournival in his poem *De vetula,* presumably written between 1220 and 1250.[5] De Fournival described the 216 ways that three dice can fall—thus including all distinct sequences, or permutations. In addition, de Fournival summarized in a table the number of three-dice *sets* which may total to the sums 3 through 18 and the corresponding number of three-dice *sequences* that can produce those outcomes (see Figure 10). The possible sums are in the leftmost columns. For

3	18	Punctatura	1	Cadentia	1
4	17	Punctatura	1	Cadentiæ	3
5	16	Punctaturæ	2	Cadentiæ	6
6	15	Punctaturæ	3	Cadentiæ	10
7	14	Punctaturæ	4	Cadentiæ	15
8	13	Punctaturæ	5	Cadentiæ	21
9	12	Punctaturæ	6	Cadentiæ	25
10	11	Punctaturæ	6	Cadentiæ	27

FIGURE 10 Richard de Fournival's thirteenth-century summary of the 216 possible sequences for the totals showing on the throw of three dice.

example, the first line displays information about the sums of 3 and 18, while the second line displays information about the sums 4 and 17. The middle column of numbers lists the number of apparent outcomes (sets) which can result in those sums, and the final column of numbers provides the number of sequences that could have produced those sums.

A sum of 3 or a sum of 18 can each occur in only one way—(1,1,1) or (6,6,6). Hence, the first line of de Fournival's table indicates that the sum of 3 or 18 can be constructed from

only one set and only one sequence. A sum of 4 or a sum of
17 each has only one apparent outcome, but that outcome
could have been produced by any one of three equally likely
sequences. For example, a sum of 4 occurs when the outcome
looks like two/one/one, which could have resulted from a
(2,1,1) sequence, a (1,2,1) sequence, or a (1,1,2) sequence. The
sum of 17 occurs when the outcome looks like five/six/six,
which could have resulted from a (5,6,6) sequence, a (6,5,6)
sequence, or a (6,6,5) sequence. In the second line of de Fournival's table, we see that the sum of 4 or 17 can be constructed
from one set produced by any of three possible sequences. A
sum of 5 or a sum of 16 is produced by either of two apparent
outcomes, and each of the two apparent outcomes can be created by throwing any of three equally likely sequences (for a
total of six sequences), and so on.

If we construct a total for the table by adding up the number
of set outcomes in the middle column, we get 28. Since each
line of the table accounts for two different sums which can be
analyzed in the same way, the total number of set outcomes is 28
× 2, or 56. If we add up the number of equally likely sequences
in the final column, we get 108. Since there are two sums per
line, we find that the total number of sequences is 108 × 2, or
216.

The step of moving from the 56 apparent outcomes in the throw of three dice to the 216 distinct throws was important in the development of a mathematical understanding of probability. Yet, the concept was not generally understood, and the theory of probability was not born until some time later. De Fournival's enumeration was either unaccepted or unnoticed at the time.

Although divination in the Christian church using dice was a rarity, Wibold, Bishop of Cambray, circa 960 A.D., described 56 virtues, which some scholars believe corresponded to the apparent outcomes on the throw of three dice.[6] The throw of three dice might have determined which particular virtue a monk ought to practice for a period of time.

A well-known enumeration of three dice throws, *Chaunce of the Dyse,* written in the early 1400s, is a medieval poem with 56 verses. Each verse is a fortune corresponding to the 56 sets (not the 216 sequences) of possible throws of three dice. It is thought that poems like this were used for informal fortune-telling, where a roll of three dice determined which particular fortune pertained to the participant. Again, for this use only the *outcome*—the final set of numbers facing up on the three dice—was of any import.

An example of one verse for the three-dice throw of 6, 5, 3 is the following:

Mercury that disposed eloquence   **61**
Unto your birth so highly was incline
That he gave you great part of science
Passing all folkës heartës to undermine
And other matters as well define
Thus you govern your wordës in best wise
That heart may think or any tongue suffise.[7]

This verse resembles modern-day horoscopes that appear in daily newspapers: You were born with the gift of rhetoric. You have the knowledge and faculty of speech and locution. You, above all others, are skilled in the art of oratory, argument, and persuasive reasoning. You debate eloquently and have the gift of persuasion.

The first evidence of the mathematical study of chance, *Liber de ludo aleae (The Book on Games of Chance)*, was written by Girolamo Cardano around 1564 but not published until almost a hundred years later. Often described as an eccentric, Cardano was a physician, mathematics instructor, occultist, superstitious gambler, and prolific author who wrote over two hundred books and manuscripts.[8] In his book on games of chance, Cardano correctly counts the 36 possible two-dice sequences and the 216 possible three-dice sequences. This allows him to accurately compute probabilities and much more.

Girolamo Cardano was not the only Renaissance Italian to demonstrate a clear understanding of random sequences. In a short essay entitled *Thoughts about Dice Games,* written between 1613 and 1623, Galileo explained why there are 216 equiprobable outcomes in the toss of three dice. He began his essay by saying that he had been asked to explain why certain sums with three dice seem to occur with an equal number of possibilities, while dice-players knew them not to be equally likely. In particular he observed that although the sums of 9 and 10 can happen with "an equal diversity of numbers," the sum of 10 was known by gamblers to be more advantageous.[9]

Of course, what he is referring to is the fact that the sums of 9 and 10 have an equal number of set outcomes, though the two sums were not equally likely. Galileo explained why the gamblers' intuition was indeed borne out by the mathematics—by showing that the six sets of outcomes that sum to 9 can occur in 25 different sequences, while the six sets of outcomes that sum to 10 can occur in 27 different sequences. It is difficult to believe that any particular gambler, no matter how skilled, could have distinguished the slight difference between 25 chances out of 216 and 27 chances out of 216. A far more likely explanation is that this knowledge was part of gamblers' lore, accumulated from experience and passed down over centuries.

□   □   □

Without a great deal of experience or an extremely keen intuition, it is very difficult to identify the equally likely outcomes of a compound event like the throw of two, and particularly three, six-sided dice. In order to compute probabilities correctly, one must be able to measure or count all the favorable and unfavorable equally likely chances. But many situations are far too complex for one to visualize those chances.

Try to envision all possible five-card poker hands. There are 2,598,960 of them! And there are over 600,000,000,000 13-card bridge hands! Though visualizing outcomes as being both compound and sequential can help us to understand probabilities in certain situations, in others it is clearly a daunting task.

Mind Games for Gamblers

5

Such fruit cometh of the bitched bones two.

GEOFFREY CHAUCER
The Pardoner's Tale

Suppose that Moe proposes a game to Larry and Curly in which two coins are tossed simultaneously. Two heads mean Curly wins, two tails mean Larry wins, one of each means Moe wins. These may seem like rules for a fair game to Curly and Larry, but if they were to toss the coins *in sequence* and observe the outcome after each toss, Curly and Larry would begin to understand that the three players do not have equal chances. After each first toss, either Curly or Larry is always eliminated and Moe is always still in the game.

To understand why this game is not fair, let's toss the coins in sequence and then pose three subtly different questions:
➤ Which outcome is more likely: two heads, two tails, or one of each? ➤ Which is more likely: one of each (a mixture) or two of the same (a match)? ➤ Which is more likely: a head followed by another head, or a head followed by a tail?

If asked this first question, some people may think (incorrectly) that the three outcomes are equally likely—each with a probability of $\frac{1}{3}$. The celebrated eighteenth-century mathematician Jean Le Rond d'Alembert, who was one of the most influential scientists in France at that time, argued that the chances of throwing a head in two tosses was $\frac{2}{3}$.[1] D'Alembert thought the equally likely outcomes were a head on the first toss, a tail followed by a head, or a tail followed by a tail. Since two out of three results contain a head, he came up with the incorrect probability of one head in two tosses as $\frac{2}{3}$.

Many people, though reasoning incorrectly, would come up with the right answer: one of each. They think that a mixed outcome more accurately reflects the long-term behavior of the coins.[2] The error in their reasoning is easily exposed when they are asked the second question: Which is more likely, a mixture or a match? Believing that heads and tails ought to occur equally often in the short run, many people answer that in two coin tosses it is more likely to get one of each (a mixture) than to get two of the same kind (a match). This, however, is not the case.

The same fallacy often occurs in answering the third question. In the sequential toss of two coins, is it more likely to get a head followed by a head, or a head followed by a tail? Those individuals who think that a head followed by a tail is more likely once again feel that coin tosses are more likely to vary from trial to trial. But this isn't so; the correct answer is that the situations are equally likely. All three questions can be answered fairly easily if we recognize the four equally likely random sequences: HH, HT, TH, and TT. Without a list of the equally likely outcomes, it is difficult to answer any probability question.

Returning to the first question—Which is more likely: two heads, two tails, or one of each?—one of each is more likely. While it may be true that visually these three outcomes (two heads, two tails, and one of each) are the only ones we recognize, they are not equally likely. HT and TH are two instances of getting "one of each" in the 4 equally likely outcomes, HH, HT, TH, and TT. There are 2 chances out of 4 of getting one of each, or a probability of $\frac{2}{4}$, which equals $\frac{1}{2}$, while two heads (HH) and two tails (TT) each have only 1 chance in 4 of occurring.

In the second question—Is it more likely to get one of each (a mixture) or to get two of the same kind (a match)?—neither is more likely; they are equally likely. The probability of getting one of each (HT or TH) is $\frac{2}{4}$, or $\frac{1}{2}$. The probability of getting two of the same kind (HH or TT) is also $\frac{2}{4}$, or $\frac{1}{2}$.

In the third question—Is it more likely to get a head followed by a head or a head followed by a tail?—they are equally likely. A head followed by a head (HH) or a head followed by a

tail (HT) implies *sequence,* and each represents exactly one instance of the four possible equally likely sequences. Therefore
each has 1 chance out of 4 (or a probability of $\frac{1}{4}$) of occurring.

A fair game (unlike the one that Moe proposed to Larry and
Curly) is one in which players have an equal opportunity to win
the same sum. For example, suppose two players bet on whether
or not a head comes up on the single toss of a fair coin. If each
player wagers $5 (in other words, each pays $5 for the privilege
of playing) and the winner collects $10, then each player has an
equal opportunity to win $5 (remember each had to pay $5 to
play). If they played the game a number of times, each player
would expect to lose $5 in about half of the games and win $5
in about half of the games.

If the players have *unequal* probabilities of winning a sum,
however, then the wager must be adjusted in order to play a fair
game. Today we frequently see this type of wagering in sports
events. If the bettors agree that two teams are of equal ability
(equally likely to win the game), then the bettors will wager the
same amounts, or bet "even money." If the bettors feel that one
team has an advantage, the wager will be adjusted. The bettor
with the superior team will be required to stake more money for
the privilege of acquiring this advantage (a higher probability of
winning the sum).

Suppose two players, Betty and Lulu, wish to play a game

which gives Betty a $\frac{7}{10}$ probability of winning and Lulu a $\frac{3}{10}$ probability of winning. Clearly if they wager the same amount, it is not a fair game. For the game to be a fair one, Betty must pay more for the privilege of playing because she has a higher likelihood of winning. If Betty pays $7 to play and Lulu pays $3 to play, then the game is a fair one. Three tenths of the time Betty will lose $7, and seven tenths of the time she will win $3. Since she is more likely to win, Betty isn't afforded the opportunity to collect as much as Lulu. The reverse win/loss situation is similarly true for Lulu.

Let us suppose the players don't know all this, and they determine their own wagers. Lulu is willing to play the game with a wager of $10 if Betty wagers $20. If they play the game many times, how much money can Betty expect to win? Let's calculate Betty's expected average winnings (her expected winnings per game). Since $\frac{3}{10}$ of the time Betty will lose $20 and $\frac{7}{10}$ of the time she will win $10, her expected per game winnings are $(\frac{3}{10} \times (-\$20)) + (\frac{7}{10} \times (\$10)) = +\$1$. If they play the game many times, Betty can expect to win an average of $1 per game and Lulu can expect to lose an average of $1 per game. In fact, if the wager were adjusted by one dollar—Betty pays $21 to Lulu's $9—the game would be a fair one.

The question of how much money players should pay in

order to play a given game was a common type of probability question in earlier times. The French naturalist Georges Louis Leclerc de Buffon, a pioneer in the use of experimentation to resolve questions of probability, described in his 1777 *Essai d'arithmetique morale* an experiment involving an empirical test of the Petersburg problem—a gambling paradox which was widely discussed by mathematicians during the eighteenth century.[3] This is the Petersburg problem:

Peter and Paul agree to play a game based on the toss of a coin. If a head is thrown on the first toss (probability = $\frac{1}{2}$), Paul will pay Peter one shilling and the game is over. If the first toss is a tail but the second toss is a head (probability = $\frac{1}{2} \times \frac{1}{2}$, or $\frac{1}{4}$), Paul will pay Peter two shillings and the game ends. If the first head appears on the third toss (probability = $\frac{1}{2} \times \frac{1}{2} \times \frac{1}{2}$, or $\frac{1}{8}$), Paul will pay Peter four shillings, and so on. Each time the winning head does not appear, the payoff doubles. Once the winning head appears, the game is over and the player is paid off. The probability that the game is won on the nth coin toss is $\frac{1}{2} \times \frac{1}{2} \times \ldots \times \frac{1}{2}$, multiplied n times, or $(\frac{1}{2})^n$. Peter is the only player who can win, so he must pay to play the game. The question is: How much should Peter pay Paul for the privilege of playing this game?[4]

If we find out what his expected winnings would be irrespec-

tive of his wager, then this is the amount Peter should pay Paul
in order to play a fair game. Since Peter can win 1 shilling
one-half of the time, 2 shillings one-fourth of the time, 4 shillings one-eighth of the time, 8 shillings one-sixteenth of the
time, and so on, we calculate his expected winnings to be:

$$(1 \times \tfrac{1}{2}) + (2 \times \tfrac{1}{4}) + (4 \times \tfrac{1}{8}) + (8 \times \tfrac{1}{16}) + \ldots$$

$$= (\tfrac{1}{2}) + (\tfrac{2}{4}) + (\tfrac{4}{8}) + (\tfrac{8}{16}) + \ldots$$

$$= (\tfrac{1}{2}) + (\tfrac{1}{2}) + (\tfrac{1}{2}) + (\tfrac{1}{2}) + \ldots$$

If we add $\tfrac{1}{2}$ an infinite number of times, this is an infinite
number of shillings. Since Peter's expected winnings are an infinite number of shillings, Peter must pay Paul an infinite number of shillings to play this game. The paradox is that, in fact,
Peter is not likely to win a large number of shillings, but a rather
small amount.

The Petersburg paradox posed a simple game of heads or
tails in which the expected winnings for one of the players,
when computed mathematically, would be infinite—yet common sense at the time indicated a modest payoff. For example,
the probability that a single game would go on beyond 10 tosses
of a coin (where the winnings would come to 1024 shillings or
more) is about 0.000977, or less than one-tenth of one percent.

In an attempt to reconcile the dilemma created by the Petersburg problem, Buffon reported that he actually performed an experiment playing the game of Peter and Paul. Using a child to toss a coin, he documented the results of 2048 games. Buffon found the winnings to average about 5 shillings per game—a modest amount. Although the expected per game winnings increase if more than 2048 games are played, Buffon points out that to increase the average winnings to 10 shillings per game, play would have to continue for 13 years![5]

The paradox can be resolved if the game does not have an indeterminate ending point. The game ends when a head is tossed, so theoretically the game could go on forever if a head is never tossed. Thus the expected winnings are also infinite. Others have said that this indicates that Paul is a liar, since he has promised to pay off and he cannot possibly cover an infinite amount even if Peter had an infinite amount to pay to play.[6]

Paul is the "house" and must insure the payoff no matter how long the game goes. Let's say that Paul does have a lot of money, so that Peter's payoff can be insured up to some very large amount. If Paul had a trillion dollars—or 10^{12} dollars—how many tosses would it take to clean out the house? The payoff for 40 tosses would just exceed (by 99 billion) a trillion dollars. What is Peter's expected return on the game if he can play at most 40 tosses (in other words Paul must pay Peter the trillion dollars if they reach 40 tosses without a head and the game will be over). By our previous formula, we now have

$$(\tfrac{1}{2}) + (\tfrac{1}{2}) + (\tfrac{1}{2}) + (\tfrac{1}{2}) + \ldots + (\tfrac{1}{2})$$

or $40 \times \tfrac{1}{2}$. So, Peter must pay $20 to play the game, Paul must have a trillion dollars to cover the payoff—just in case the game goes to 40 tosses—and there is close to a 97 percent chance that Peter will win less than $20.

Equally Likely in the Long Run

The headline of a 1990 *New York Times* article reads as follows: "1-in-a-Trillion Coincidence, You Say? Not Really, Experts Find."[7] The article goes on to report a seemingly unbelievable coincidence about a woman who won the New Jersey lottery twice within four months, a feat originally reported as a 1 in 17 trillion long shot. Research on coincidences by two Harvard statisticians revealed, however, that the odds of such an event happening to *someone somewhere* in the United States were more like 1 in 30—not that amazing after all. They explain that this is an example of the law of very large numbers: "With a large enough sample, any outrageous thing is likely to happen."[8] Out of the millions upon millions of people who regularly purchase lottery tickets in the United States, it is not unreasonable that someone should at some point hit the lottery twice.

Another article in the *New York Times* in September 1996, this time regarding the tragic TWA Flight 800 crash on July 17, 1996, reported that "more than once, senior crash investigators have tried to end the speculation by ranking the possibility of friendly fire at about the same level as that a meteorite destroyed the jet." In a letter to the editor on September 19, Charles Hailey and David Helfand wrote: "The odds of a meteor striking TWA Flight 800 or any other single airline flight are indeed small. However, the relevant calculation is not the likelihood of any particular aircraft being hit, but the probability that one commercial airliner over the last 30 years of high-volume air travel would be struck by an incoming meteor with sufficient energy to cripple the plane or cause an explosion." Noting that 3000 meteors with the requisite mass hit the earth every day, and that 50,000 commercial airline flights occur worldwide each day; and assuming an average flight time of two hours, which translates into more than 3500 planes in the air at any moment; and calculating that these planes would cover approximately two-billionths of the earth's surface, the authors conclude that, in over 30 years of air travel, the probability that a commercial flight would have been hit by meteoric impact sufficient to crash the plane is 1 in 10. If their assumptions and calculations are accurate, this would be another instance of the law of very large numbers.

In the case of the Petersburg problem, the chances that Peter will toss 40 tails in a row in a given game are exceedingly small. But if enough games are played, eventually someone will win the trillion dollars. In the long run, even the most unlikely things do happen.

Cicero, who had a surprisingly modern understanding of probability, understood this well in the first century B.C. Faced with the statement that 100 consecutive Venus-throws could not occur by chance, Cicero disagreed:

> You said, for example, "For the Venus-throw to result from one cast of the four dice [*talis*] might be due to chance; but if a hundred Venus-throws resulted from one hundred casts this could not be due to chance." In the first place I do not know why it could not . . . Nothing is so uncertain as a cast of dice and yet there is no one who plays often who does not sometimes make a Venus-throw and occasionally twice or thrice in succession. Then are we, like fools, to prefer to say that it happened by the direction of Venus rather than by chance?[9]

Unlike his contemporaries who felt that these things happened at the direction of the gods, Cicero's more mature under-

standing was that *any* throw, and any sequence of throws, might   **75**
be attributed to chance. Cicero knew that in the long haul,
given enough opportunities, the rare event will occur. This
shrewd observation is not the least bit obvious to most people
even today.

Rabbi Isaac ben Mosheh Aramah, writing in the fifteenth
century, held a different view about an event which occurs many
times in a particular way. He considered it a miracle. Comment-
ing on Scripture which describes the use of lots to identify Jonah
as the one guilty of bringing a great storm, he says:

> For it is impossible for it to be otherwise than that the
> lot should fall on one of them whether he be innocent
> or guilty . . . However the meaning of their statement
> "let us cast lots" is to cast lots *many times*. Therefore the
> plural—*goralot*—is used rather than [the singular] . . .
> They did so and *cast lots many times* and every time
> the lot fell on Jonah and consequently the matter was
> verified for them. It follows then that the casting of *a* lot
> indicates primarily a reference to chance.[10]

The cast of a single lot is considered to be chance, whereas a
cast many times repeated is considered to be a sign of God and
therefore not chance. In a similar fashion, the Rabbi remarks on
the role of chance in the selection of lots for the scapegoat, again

emphasizing that the achievement of the good omen of the lot "for the Lord" in the right hand ought only to be considered a miracle if it happens many times.

Rabbi Aramah indicates that ordinary lots (as opposed to those behaving at divine direction) have no tendency to one side over the other; in other words, each lot is equally likely to be chosen. His definition of a miracle or a "sign" excludes any one-time occurrence, which could happen by chance. A string of repeated events, however, may qualify as a miracle, for such, he implies, would be unlikely to occur by chance. Cicero would not have agreed.

Writing around 1564, Girolamo Cardano refers to a certain unlikely dice throw and assures us that "still in an infinite number of throws, it is almost necessary for it to happen." Cardano is emphasizing that with *many* trials the number of times a particular outcome will occur is very close to mathematical conjecture, or mathematical expectation. And *this applies to even the most unlikely events; if it is possible for them to happen, given enough opportunities, eventually they will happen,* in accordance with the laws of probability.

Cardano, like Cicero, understood that the rare event will occur in the long run. In his *Book on Games of Chance,* Cardano asserts: "The most fundamental principle of all in gambling is simply *equal conditions,* e.g. of opponents, of bystanders, of money, of situation, of the dice box, and of the die itself . . . For in play with dice you have no certain sign, but everything depends entirely on *pure chance,* if the *die is honest.* Whatever there may be in it beyond unfounded conjecture and the arguments given above should be put down to *blind chance.*" Here, Cardano defines the fundamental underpinnings of gambling and chance as requiring equal conditions.[11]

Cardano goes on to exhibit an understanding of equiprobable alternatives, emphasizing the difference between one particular random outcome and what we would experience over a long period of time: "The talus has four faces, and thus also four points. But the die has six; in six casts each point should turn up once; but since some will be repeated, it follows that others will not turn up." Illustrating the random nature of chance events, Cardano points out that in six casts of a die a particular side may come up or not, but in a large number of trials each side of a die will face up 1 time in 6.[12]

For example, in 300 casts of a single die, we can expect a one to face upward about 50 times, since 50 out of 300 is equivalent to 1 in 6—our mathematical expectation. In 30 casts we can expect a one about 5 times. Each expectation is based on the probability of getting a one 1 time in 6, or $\frac{1}{6}$ of the time. In actuality, we may or may not get a one $\frac{1}{6}$ of the time, but the larger the number of trials, the closer the proportion of ones will be to $\frac{1}{6}$.

In 10,000 rolls of a die, mathematical expectation tells us that a one will appear approximately 1667 times, since 1667 is about $\frac{1}{6}$ of 10,000. We might actually count 1660 ones, or 1690 ones, or 1700 ones but the *proportions* of ones—1660 out of 10,000, 1690 out of 10,000, and 1700 out of 10,000—are very close to 1667 out of 10,000. Converted to percentages, they are 16.6, 16.9, and 17.0 percent, respectively, which are all close to the 16.7 percent predicted by the mathematics of probability. In 100,000 rolls of a die, we could actually count 500 more ones than predicted mathematically and still the percentage of ones would be affected by only one-half of 1 percent.

By the ages of 7–11, children notice that the larger the number of trials, the closer the results are to mathematical expectation. By ages 11–12 years this intuition often solidifies into an understanding of probability based on long-run frequencies. At very young ages children do not understand this concept. Part of the problem is that young children do not accept the notion of randomness, which is at the heart of any understanding of probability. Piaget and Inhelder found that young children conceive of random results as displaying regulated but hidden rules. When asked to predict which of several colors a

spinner would stop on, they consistently gave one of two responses: either the spinner ought to return to a color previously
landed on, or—just the opposite—the spinner ought to land on
a color not yet landed on.[13]

Yet the views of many adults expressed at the gaming table
sound much like those of young children. After a string of wins,
we may overhear, "He's hot, bet with him," or after a string of
losses, "He's due, bet with him."

After a large number of losses, gamblers often think, "My
luck's got to change," or "Soon the odds will be in *my* favor."
The psychologists Daniel Kahneman and Amos Tversky point
out that the heart of the gambler's fallacy lies in a misconception
about the fairness of the laws of chance. We believe chance to be
a self-correcting process—in which deviations in one direction
will soon be countered with a deviation in the other direction.
But in fact, deviations in the short run are not corrected; they
are merely diluted over the long run, as Tversky and Kahneman
point out.[14]

In her weekly column in *Parade Magazine,* the enormously
popular Marilyn vos Savant, purported by the *Guinness Book of
World Records* to have the world's highest IQ, is asked the following question by a reader, "Somehow you've overcome extreme
odds and flipped 100 consecutive heads. The chances of flipping
another head on the next toss can't possibly be as great as 50-50,

can they?"[15] Indeed, many people feel as this reader does, that 100 heads is sure to be soon offset by a tail.

But as the reader herself points out, a person who flips 100 heads in a row has already overcome extreme odds to get where she is (Figure 11). The question now is only about one toss. If the coin is fair, her chances are indeed 50-50, the same as they were on the first toss. But perhaps the 100 previous tosses present some evidence that the coin is not fair—it might, in fact, be biased toward heads. In that case, the chances of a head on the next toss are better than 50-50.

Games using dice, coins, and other chance devices had been popular in Europe for over 2000 years. Yet it was not until the late sixteenth and early seventeenth centuries in Italy that learned men developed a solid understanding of probability. While gamblers had by that time gained considerable expertise regarding games of chance, only the writings of Cardano and Galileo appear to exhibit a deeper insight into the mathematics of probability.

It is ironic, though not surprising, that the first mathematical writings about chance were prompted by those with the most experience observing chance and "luck"—dice-throwing gamblers. Cardano, who had an impressive understanding of probability for the times, was himself a superstitious gambler, and

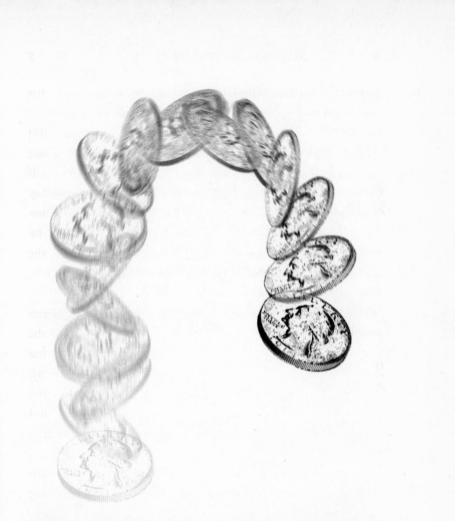

FIGURE 11 One hundred heads in a row. Is it luck? Is it a miracle? Is the coin rigged? Or am I on a roll?

Galileo's small but important tract *Thoughts about Dice Games* was written for the gambling noblemen at the court in Florence.[16]

The beginning of the serious study of probability is usually attributed to the correspondence between Blaise Pascal and Pierre de Fermat, which began around 1654. This correspondence also originated in a gambling problem related to Pascal by his friend the Chevalier de Mere, a prominent figure at the court in Paris and a gambler.[17] Fortunately, because Pascal and Fermat were considered mathematicians of the highest caliber, their interest in probability drew the attention of other mathematicians and philosophers to the serious study of the laws of chance.

Chance or Necessity?

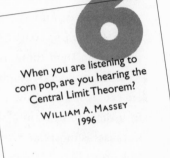

When you are listening to corn pop, are you hearing the Central Limit Theorem?

WILLIAM A. MASSEY
1996

Is a random outcome completely determined, and random only by virtue of our ignorance of the most minute contributing factors? Or are the contributing factors unknowable, and therefore render as random an outcome that can never be determined? Are seemingly random events merely the result of fluctuations superimposed on a determinate system, masking its predictability, or is there some disorderliness built into the system itself?

The philosophical question of chance versus necessity was raised as far back as the Greeks, and it is still being debated today. The first atomist, Leucippus (circa 450 B.C.), said, "Nothing happens at random; everything happens out of reason and by necessity." The atomic school contended that chance could not mean *uncaused,* since everything is caused. Chance must instead mean *hidden cause.* This view was carried on by

Leucippus' successor, Democritus (circa 460–370 B.C.), for whom chance meant *ignorance* of the determining cause.[1]

In later centuries, the Stoics also repudiated the causeless event. Chrysippus (circa 280–207 B.C.) is quoted as saying, "For there is no such thing as lack of cause, or spontaneity. In the so-called accidental impulses which some have invented, there are causes hidden from our sight which determine the impulse in a definite direction."[2] The reference to so-called accidental impulses is most likely an allusion to the contrary philosophy of Epicurus (341–270 B.C.) and his disciples.

The Epicureans held that the belief in necessity—that is, that events are preordained—and the elimination of uncaused events were incompatible with the concept of free will. Epicurus accepted the basic theory of the atomists, but departed from them in believing that a spontaneous, uncaused swerve of the atoms caused them to collide, and the uncertainty of the swerve admitted the element of free will. The Epicurean philosophy is best documented by Lucretius, the Roman poet who lived from 96 to 55 B.C., in his poem *De rerum natura*. Whereas chance for the atomists Leucippus and Democritus meant ignorance of cause, for Epicurus and Lucretius chance encompassed the indeterministic.

A deterministic worldview was predominant during the Middle Ages in Europe, encouraged by the early Christian church's belief that everything happened at the behest of the Creator. The English philosopher Thomas Hobbes (1588–1679) held that all events were predetermined by God or by extrinsic causes determined by God. His universe had no place for chance; everything proceeded from necessity. In a debate between Hobbes and Dr. Bramhall, the Bishop of Derry, documented around 1646, Hobbes stated that it is ignorance of those causes that prevents some men from perceiving the necessity of events and attributing them instead to chance.[3]

The Bishop, however, had different ideas. While he rarely used the word chance in his attempt to argue that man does have free will, he admitted to an indeterminacy in events which he called *contingent*. Contingent events are events which may or may not happen, or may happen in some other way. As an example, the Bishop used the throw of two dice which results in a pair of ones (sometimes called "snake eyes"):

> Supposing the [position] of the parties' hand who did throw the dice, supposing the figure of the table and of the dice themselves, supposing the measure of force applied, and supposing all other things which did concur to the production of that cast, to be the very same they were, there is no doubt but in this case the cast is necessary. But still this is but a necessity of supposition; for if

all these concurrent causes, or some of them, were contingent or free, then the cast was not absolutely necessary.
To begin with the caster, he might have denied his concurrence, and not have cast at all; he might have sus-
pended his concurrence, and not have cast so soon; he
might have doubled or diminished his force in casting,
if it had pleased him; he might have thrown the dice
into the other table. In all these cases what becomes of
his *ambs-ace* [pair of ones]? The like uncertainties offer
themselves for the maker of the tables, and for the maker
of the dice, and for the keeper of the tables, and for the
kind of wood, and I know not how many other circumstances.

The Bishop was pointing out that although any particular throw
is determined by physical law, there are so many uncertain aspects to the set-up that the outcome is indeterminate.[4]

Hobbes countered that if a single aspect of the set-up is
altered, a different throw is necessarily determined, and that
indeterminacy simply means we do not know which necessity
will occur. The Bishop retorted that the effect is necessary "when
the cast is thrown; but not before the cast was thrown." The
Bishop was making a most astute distinction: the physics of the
chance set-up is deterministic when it happens but indetermin-

istic before it happens. This was not the generally accepted view
at that time.[5]

If all the initial conditions were known, would the toss of
dice be random? The Bishop of Derry, circa 1646, probably
would have said that if the initial conditions can be completely
known or completely controlled, then the toss of dice is not in
the realm of chance events. If one could build a machine that
could produce a certain dice throw with 99 percent accuracy,
then the Bishop would probably say that it was 1 percent contingent.

Today, we would still allow that the throw from the machine was random; the outcomes are simply not equally likely.
And even if we could increase the accuracy of the machine to
99.9 percent, we would still consider the throws as random—
each with a certain likelihood of occurring. But once the accuracy becomes 100 percent, or the outcome certain, the event is
purely deterministic and no uncertainty is involved. No uncertainty—no randomness.

A short time after the debate between the Bishop of Derry
and Thomas Hobbes, a scientific discovery was made which,
oddly enough, complemented Christian determinism and, for
the next two centuries, weighted the debate in favor of necessity
over chance. That discovery was Newtonian physics—a system
of thought which represented the full bloom of the Scientific

Revolution in the late seventeenth century. Based on this work of Newton and many others, a belief developed among scientists that everything about the natural world was knowable through mathematics. And if everything conformed to the design of mathematics, then a Grand Designer must exist. Pure chance or randomness had no place in this philosophy.

This determinism can be seen clearly in John Arbuthnot's 1692 preface to the translation of Christiaan Huygens' *De retiociniis in ludo alea (Calculating in Games of Chance),* the first printed text on the theory of probability. Arbuthnot states, "There are very few things which we know, which are not capable of being reduc'd to a Mathematical Reasoning, and when they cannot, it's a sign our Knowledge of them is very small and confus'd." Concerning chance, he said, "It is impossible for a Dye, with such determin'd force and direction, not to fall on such a determin'd side, only I don't know the force and direction which makes it fall on such a determin'd side, and therefore I call that Chance which is nothing but want of Art."[6]

As scientists set about diligently investigating the earth and the heavens around them during the Scientific Revolution, they attempted to reduce everything to mathematics. In the fervent belief that exact numerical analysis must lead to universal laws, they measured distances on earth, distances in space, orbits, tides—everything, it seemed, was within the measuring, counting, and calculating ability of man.[7]

But despite the finest instruments and methods, capable scientists continually came up with different measurements when measuring the same entity. Even the same scientist, taking more than one observation of the same thing, was likely to get different results. Random error was turning up, quite unexpectedly, in the works of natural scientists intent on discovering facts about *exact* phenomena, such as astronomical distances and orbital paths—phenomena which appear to represent the antithesis of chance.

The Theory of Errors

Scientists trained in the disciplines of astronomy and geodesy (the study of the size and shape of the earth) were forced to grapple with the problem of how to account for discrepant measurements of a given phenomenon that was presumed to be constant. A great deal of discussion arose about how to deal with the discrepancies caused by imperfect instruments or imperfect observers. If all measurements resulted in slight differences, which measurement was the true measurement? How much of a difference is "slight"? If a new measurement was taken—one for which there were no other comparison measurements—how likely is it to be off by a slight amount or a great amount? The amount by which an individual observation departed from the true measurement (the difference between the two) was referred to as an observation error, or measurement error.[8]

As early as the late sixteenth century, the Danish astronomer Tycho Brahe was among the investigators who strived to eliminate random errors in order to achieve exactness in scientific measurement. By 1632 Galileo had formulated several propositions about observational errors: (1) that errors are unavoidable, (2) that small errors are more likely than large ones, (3) that measurement errors are symmetrical (equally inclined to overestimation as to underestimation), and (4) that the true value of the observed constant is in the vicinity of the greatest concentration of measurements.[9]

Since any one observation was subject to random error in measurement, by the eighteenth century scientists were greatly interested in the *combination* of several observations. One way to combine observations is by averaging them. Indeed, by the first half of the eighteenth century the majority of investigators considered the arithmetic mean (the average) as the most accurate representation of the "true" value of discrepant measurements.[10] The mean, or average, would not be as accurate as the best measurement nor as inaccurate as the worst measurement.

But since no one knew which was the best and which was the
worst, the mean was at least a safe choice.

Taking the average of several observations was thought to be a reasonable estimate of the exact value being measured. For example, suppose we measure the length of a sheet of paper with highly sophisticated instruments and obtain five different results: 11.05, 11.09, 10.96, 11.01, and 10.98 inches. Any one of these could be the correct measurement, or none of them could be correct. Certainly one of them is the most accurate (or more than one in the case of a tie), and one (or more) is the least accurate. If we need one number to use as the length, it seems reasonable to choose a number which can represent the middle of the group of numbers, since it cannot be as extreme as the highest or lowest observation.

In an essay written in 1756, Thomas Simpson attempted to demonstrate the advantage of using the mean over a single observation in astronomy. He showed that the chances of the mean being in error by certain amounts were less than the chances of a single observation being in error by the same amount. Simpson compared random errors in measurement to the throw of dice, and for this very important insight he is considered by many to be the first person to associate the distribution of errors with chance events.[11]

In his analysis, Simpson assumed that the probability distribution of errors in a single measurement was analogous to the probability distribution of the sums of dice throws. Let's examine how he might have arrived at this supposition. If an observation is equally likely to be any number within a reasonable range of the true measure, then the probabilities of any single observation are distributed uniformly. The chances of a particular sized error (the difference between the actual observation and the true measurement) would also be uniformly distributed—as are the probabilities of a number showing on the throw of a single die (see Figure 12, top).

But if we look at the way a group of errors is distributed, we will not see uniform probabilities. More errors will cluster near the middle of the group, with the number of errors tapering off as the size of the error is further from the middle, until there are very few at the extremes (Figure 13, top). Notice how similar Simpson's distribution is to the probability distribution of dice sums when *more* than one die is rolled (Figure 12, bottom). There, the likelihood of sums near the middle is great, and the likelihood of sums decreases away from the middle.

Simpson hypothesized that the chances of errors of observation in astronomy were proportional to their size and would thus behave in a manner similar to the probabilities in dice sums. He thereby arrived at a triangular shape to represent the probability distribution for observational errors in astronomy.

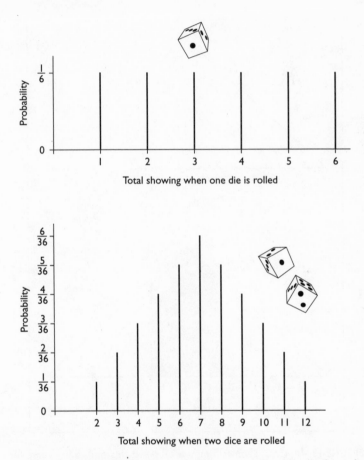

FIGURE 12 Probability distribution for the outcome on the roll of one die (top), compared with the probability distribution for the outcome, or sum, on the roll of two dice (bottom).

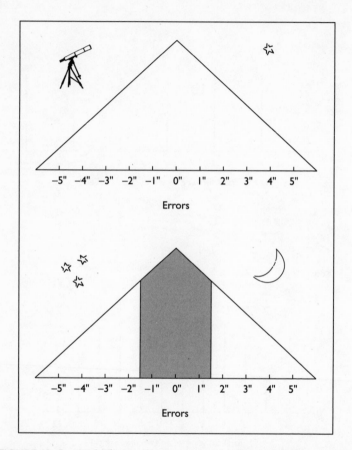

FIGURE 13 In 1756 Thomas Simpson proposed this probability curve for errors in astronomy (top), and suggested that the probability of an error between −1″ and +1″ could be found by calculating the shaded area under the probability curve (bottom).

The main difference in the appearance of Simpson's probability curve and the appearance of the probability graph for two-dice sums is attributable to the difference between the types of variables being measured. The sum on two dice can be only the numbers 2, 3, 4, 5, and so on through 12. The sum is called a *discrete* random variable, because it can vary (by chance) only among eleven specific outcomes. On the other hand, Simpson's errors can be *any* numbers between −5" and +5"—for example, an error can be 1.5", 1.51", 1.501", or any number of seconds. Observational errors are examples of a *continuous* random variable, because each observation can vary (by chance) among any of the numbers on the continuum between −5" and +5".

The probability of a certain sum on two dice can be read directly from its probability graph by observing how high the spike or bar rises. For example, in the bar graph representing the probabilities of two-dice sums, the spike above the sum of 7 rises to a height of $\frac{6}{36}$, or $\frac{1}{6}$. This indicates that the probability of obtaining a 7 on the roll of two dice is $\frac{1}{6}$. The probability of rolling one or more sums can also be read from this graph. For instance, the probability of rolling a 6, 7, or 8 is calculated by summing the heights of the spikes above 6, 7, and 8, or $\frac{5}{36} + \frac{6}{36} + \frac{5}{36} = \frac{16}{36} = \frac{4}{9}$.

By contrast, we cannot read the probabilities of various-sized errors directly from Simpson's graph. The probabilities for errors (and for any continuous variable) are read by calculating areas under the probability curve. For example, the calculation for the probability of obtaining an error within one second of the true measure involves computing the probability of all possible errors from −1" to +1" (the shaded region in Figure 13, bottom). Simpson calculates this probability to be approximately 0.44. Notice that the area actually calculated is not from −1" to +1" but rather from −1.5" to +1.5". This is because Simpson presumed the observational errors to be to the nearest second, and therefore a measurement that turned out to be 1.47", for example, would be rounded to 1".

In an essay written in 1777, Daniel Bernoulli took exception to the idea that all errors are equally likely. Bernoulli likened the chance errors of astronomical observation to the deviations of an archer.[12] In describing the deviations from a target along a vertical line, he comments, "All the errors are such as may easily be in one direction as the other, and their outcome is quite uncertain, being decided only as it were by unavoidable chance." But he adds, "Is it not self-evident that the hits must be assumed to be thicker and more numerous on any given band the nearer this is to the mark?"

Bernoulli illustrated his reasoning by pointing out the difference in choosing the "most probable" throw from among equally likely dice throws, as opposed to choosing the "most probable" throw when dice throws are not equally likely. Daniel Bernoulli clearly saw the errors in astronomical measurements as random outcomes, adding that "the whole complex of observations is simply a chance event." But he also thought that the probability curve for observational errors would be semicircular, with the length of the radius being the greatest error one would likely make (see Figure 14, top).

Origins of the Bell Curve

In a 1778 memoir Pierre-Simon, Marquis de Laplace, discovered a probability curve for sums or means that would become *the* curve for errors of observation (see Figure 14, bottom). Later, in 1808, the normal curve for the distribution of random errors, as this bell-shaped curve is formally described, was developed independently by an obscure American, Robert Adrian, and one year later by the famous German mathematician, physicist, and astronomer Carl Friedrich Gauss. Adrian derived the normal curve formula as the *curve of probability* for different errors of observation in navigation, astronomy, and geodesic survey. Gauss derived the normal curve as the law that described the probability of errors in astronomical observations in a study on the motion of heavenly bodies. In fact, the normal distribution of errors is sometimes referred to as the Gaussian distribution.[13]

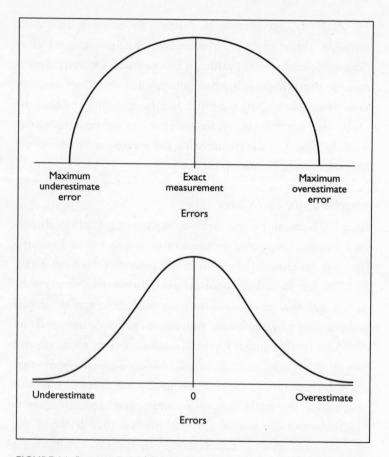

FIGURE 14 In 1777 Daniel Bernoulli suggested a semicircular shape for the probability curve for errors in astronomy. One year later, the normal probability curve for errors of observation, or bell curve (bottom), was proposed by Pierre-Simon de Laplace.

In an 1810 memoir read to the Academy in Paris, Laplace   **99**
presented what may be his major result in probability theory,
now known as the Central Limit Theorem. By 1812 his theorem was fully developed, as Laplace had by then made the
connection between Gauss's work using the normal curve as the
distribution curve for errors and his own discovery of the normal distribution for sums or means of random events. The
Central Limit Theorem proved that the sum or mean of a great
number of errors would be approximately normally distributed.[14] Since observational errors were presumed to behave like
simple chance events, we can restate the theorem to say that the
sum or mean of a great number of *independent random observations* is approximately normally distributed.

The Laplace/Gauss synthesis put forward the idea that because certain phenomena, such as errors, behaved like chance
events, they could be predictably described by probability distributions—particularly by the normal or bell-shaped distribution.[15] Moreover, not only do the frequencies of certain data,
like random errors, follow a normal distribution but the probability distribution of the sum or mean of any such data will be
approximately normal.

Thus, the normal distribution curve began as the *theory of
errors* in disciplines where errors of measurement or fluctuations
of nature were believed to behave randomly. During the next ten

to twenty years, in studies of astronomy, physics, and even artillery fire, the Central Limit Theorem came to be considered a universal law—the normal law of random errors.[16]

Prior to the discovery of the bell curve as the probability curve for errors in measurement, its formula was derived by Abraham De Moivre for an entirely different purpose: to estimate discrete probabilities, particularly ones that involved laborious calculations. His discovery was published in a supplement to the 1733 edition of *The Doctrine of Chances.*[17] De Moivre was writing about events that have the same probability of happening or not happening—similar to the toss of a coin where a head is just as likely to face up as not. In looking at a large number of such events, De Moivre was interested in computing the probabilities for the total number of occurrences of a particular outcome. For instance, the question might involve computing the probabilities for the total number of heads when a coin has been tossed many times. Let's examine this question and see what happens as the number of trials becomes very large.

In one coin toss, we know that we can get a total of 0 heads (when a tail faces up) or 1 head. Since the outcomes of 0 heads or 1 head are equally likely, the probability bar graph shows spikes of equal height, indicating the equally likely probabilities. When a coin is tossed twice, we can get a total of 0 heads, 1 head, or 2 heads. The probability of 1 head is twice as likely as either 0 heads or 2 heads (see Figure 15).

As the number of tosses increases, the overall shape of the corresponding probability distribution changes. There are more possible outcomes, the probabilities are not as great for any one outcome, and therefore the spikes are not as high. If we consider even more tosses, this pattern will continue. Consequently, more and more outcomes will be possible, and even the highest spike will not be very high. With twenty-five tosses the highest spike indicates a probability of less than 0.155.

Though it was known during De Moivre's time how to compute these probabilities, the calculations themselves proved to be quite tedious when worked out by hand. De Moivre discovered that a very good approximation of these probabilities could be obtained by using an entirely different method—computing areas under what is now referred to as the normal curve. De Moivre deserves credit for first discovering the normal curve formula, though he developed it as a way of estimating discrete probabilities and not as a way of coping with errors in measurement.[18]

Let's look at one more example that will highlight the connection between De Moivre's work and Laplace's discovery about the probability of sums being normally distributed as the

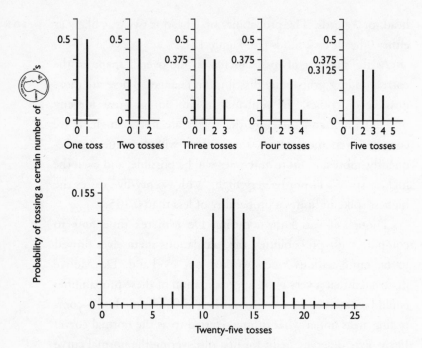

FIGURE 15 Probability graphs for the total number of heads showing when a coin is tossed one, two, three, four, five, and twenty-five times.

number of observations increases. When one die is rolled, the **103**
probabilities of the six outcomes (1 through 6) are equally likely,
and the bar graph of those probabilities indicates spikes of uniform height (see Figure 16). When two dice are rolled, the graph
of the probabilities of the eleven sums (2 through 12) shows
spikes that take on the shape of a triangle. When three dice are
rolled, the sixteen possible sums are 3 through 18, and the
overall pattern of the spikes changes again. With each increase
in the number of dice, the vertical scale (the probability) gets
shorter and the horizontal scale (the possible sums) gets wider.
When four dice are rolled, 21 sums (4 through 24) are possible,
and the shape of the distribution of sums is beginning to look
like the familiar bell curve.

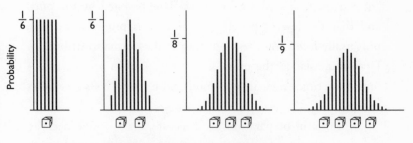

Total showing for one, two, three, and four dice

FIGURE 16 Probability graphs for the sums showing on the roll of one, two, three, and four dice.

In 1873–74 Sir Francis Galton (Charles Darwin's cousin) designed an apparatus that he later named the quincunx.[19] This machine was a clever physical model of the theory of errors, which he believed was applicable to many phenomena in biology as well as physics. Enclosed behind glass was a cross section of a funnel opening onto a pyramid of equally spaced pins, with vertical compartments below the pins. Pellets, once released from the funnel, would bounce helter-skelter, left or right, against the pins (representing, in Galton's theory, the independent random disturbances of nature), to ultimately gather in the lower compartments in a pile which resembles a normal curve (see Figure 17). Galton called this phenomenon the law of deviation. Galton believed that the important influences that acted upon an inherited characteristic, such as height, were a "host of petty disturbing influences" (represented by the pins) and that the law of genetic deviation was purely numerical, universally following the single law of the normal distribution. Devices similar to the quincunx can be seen in science museums; sometimes they are enormous, with tennis balls serving as pellets.

In 1877, in preparation for a lecture, Galton modified the quincunx to represent a corollary to this theory: namely, that variability in inherited characteristics is offset by reversion to the mean. The two factors of variability and reversion, when taken

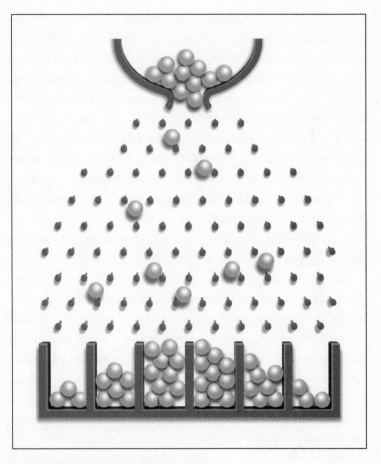

FIGURE 17 Francis Galton's early quincunx.

in aggregate, tend to produce a generation which resembles the previous generation. To illustrate these effects within the inherited characteristics of families, Galton added a second level to the quincunx, which represented a generation of offspring (see Figure 18). Once the pellets rested in compartments in a pile resembling a normal distribution, a trap door below one local compartment was opened and those pellets were allowed to continue through a second maze of equally spaced pins. Those pellets also landed in a small pile, resembling a normal distribution. If every trap door was opened, many small piles would form and the whole distribution of pellets would resemble the original pile—demonstrating the way in which generations of offspring tend to resemble the generation of parents.[20]

In the seventeenth and eighteenth centuries, as the theory of probability evolved slowly from games of chance and gained momentum among mathematicians whose interest was piqued by problems in gambling, the field of statistics was advancing along a seemingly unrelated front—in astronomy and geodesy, in biology, and in actuarial work. As the application of statistics to the natural sciences proceeded, both physicists and biologists adopted the astronomers' theory of errors as their own.

During the nineteenth century, an enormous number of natural phenomena in the life sciences, as well as in physics,

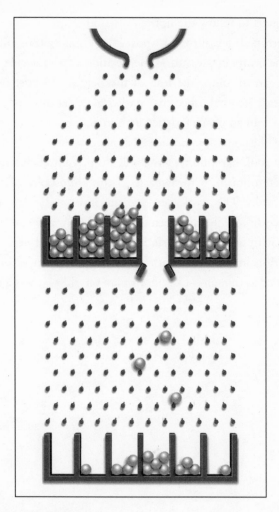

FIGURE 18 Galton's 1877 quincunx.

were found to follow the normal distribution of random errors. If the errors in measuring the position of a star were random, so were the errors in measuring the position of a molecule. Biologists began to follow the lead of the Belgian scientist Adolphe Quételet, who established that statistical deviations from the average in human characteristics such as height, weight, strength, and the like could also be viewed as errors.

Few took exception to the still deeply entrenched Newtonian determinism, a throwback to the philosophy of the ancient Greeks. This worldview saw precision, order, law, design, and necessity in the arrangement of the universe, and equated chance with ignorance. As statistical theory gained new areas of application in the nineteenth century, the bell curve began to be applied as though—like Newtonian physics—it was a universal law.

Order in
Apparent Chaos

7

Children attending a day camp list their ages to the nearest year: 6, 7, 5, 6, 6, 7, 6, 5, 7, 6, 5, 6, . . . , 7. Because no rule can determine the next number in this sequence, the sequence might be considered to be as random as the toss of a coin. Yet, the laws of probability instruct us that the distribution of sums or means of samples taken randomly from a lengthy sequence such as this one do tend to form the familiar bell curve. The idea that such a nonsystematic, nondeterministic sequence can yield predictable results is a powerful one. Indeed, in the century following the discovery of the Central Limit Theorem by Laplace and Gauss, scientists began to use (and misuse) this idea, bringing the results to a wide variety of fields.

> From where we stand the rain seems random. If we would stand somewhere else, we would see the order in it.
>
> TONY HILLERMAN
> *Coyote Waits*

The real strength of what Laplace and Gauss discovered is that inferences can be drawn from data, but only if the data are compared *against* something—and that "something" is usually the behavior of random data. The only way we can legitimately rate or compare the value of an observed sample statistic, such as a mean or sum, within the hypothesized sampling distribution is to select a sample randomly.

In trying to project the annual number of births in France from a sample of data, Laplace seems to have lacked any concept of a random sample, though he states that his data were chosen "to render the general result independent of local circumstances." Laplace and many others to follow typically treated their data as though all data had the characteristics of chance events; the randomness of their samples was merely assumed. Though the basis of the Central Limit Theorem is that observations must be taken at random, this aspect was largely ignored in practice—not just by those applying probability and statistical theory but by those attempting to expand and extend the theory as well. Until the late 1920s, most of the developers of statistical theory seemed eager to use whatever data were at hand at the moment—merely assuming that a set of data was a random slice of a larger population. What they seem not to have realized was that, even if *errors* are independent and random, not all data are.[1]

As the modern mathematical field of statistics developed, the majority of the natural and social scientists continued to apply statistical principles to whatever large amounts of data existed. Most were skeptical of random sampling, perhaps not fully understanding that it could be beneficial, not harmful, when put to inferential use. A few, however, had begun what was to eventually become standard procedure in statistical experimentation—random sampling, or the generation of random data. Experiments began with dice, counters, and slips of paper—at times to suggest or confirm a new theory, at other times to illustrate an existing theory to colleagues, and sometimes to teach statistical concepts to a broader audience.

One of the relatively few developers of statistical theory to use random samples was Gustav Theodor Fechner. He began his work in probability and statistics in Germany shortly after 1852.[2] Described as the founder of psychophysics and the first universal indeterminist, Fechner developed tests whereby variation due to factors other than chance could be detected by comparing the data under consideration to a random sequence.

To generate a random sequence, Fechner obtained the numbers drawn in ten Saxon lotteries from 1843 to 1852 (32,000–34,000 numbers each), and he reports that these digits were used in the order in which they were drawn. To these lottery sequences, he compared meteorological data, data on births, deaths, and suicides in different seasons, male and female births, and the number of thunderstorms in different places, to determine whether fluctuations in these sets of data should be credited to some local circumstance or attributed solely to chance.[3]

In the case of his meteorological data, Fechner concluded that certain patterns of weather were indeed dependent on the data preceding them, and therefore the laws of chance were "disturbed" by the laws of nature. Fechner's work was exceptional, yet it received little notice until around 1908, when the world was finally ready for its indeterministic aspects.

Another nineteenth-century instance of using a randomly chosen sample to study distribution theory is the 1876 work of an American, Erastus L. DeForest. DeForest worked for an insurance company, and much of his statistical analysis was conducted for the actuarial field. In attempting to confirm whether certain algebraic formulas (functions) were capable of modeling a mortality curve, DeForest created a method of adjustment to data which accounted for errors of observation. In the process of testing this method, he performed the first known simulation by using what we would today call artificial random data.[4]

DeForest devised an elaborate lottery-type method to obtain data representing random errors. He transcribed 100 values from a normal curve table (at equally spaced probabilities) and copied these values onto bits of cardboard of equal size. These were then shaken in a box, drawn one by one, and copied down

in exactly the order in which they were drawn. The bits of cardboard were returned to the box, shaken, drawn again, and copied down. The simulation was carried out a number of times.
The means of the expected errors for various-sized samples were
then compared to the means of the observed errors when his
formula was used. DeForest's work was not widely known to
other statisticians at the time, however, and his attempts to
ensure randomness in applying probability and statistical theory
to data were largely ignored.

In addition to Fechner and DeForest, a few other people
attempted to generate random samples on occasion. In 1877
George Darwin, son of Charles Darwin and cousin to Francis
Galton, invented a spinner to generate "errors" for adjustment
to meteorological data. Galton himself in 1890 created a set of
three dice that would generate random values from a normal
distribution. Galton mentioned that his method might be used
when statisticians wish to test the practical value of some process
like the adjusting or "smoothing" of data.[5] In 1883–84 Charles
S. Peirce and Joseph Jastrow introduced a random element into
their plan for a psychology experiment on sensation percep-
tion. A card chosen by an operator from a shuffled deck would
determine whether he ought to diminish or increase the stimu-
lus applied to the subject. Peirce and Jastrow indicated that in
choosing a card at random to determine this aspect of the ex-

periment, the decision was removed from the operator. They felt that this would prevent unintended bias by the operator or the subject.[6]

One of the first random sampling experiments in statistical literature has been attributed to Francis Ysidro Edgeworth in 1885.[7] Edgeworth's work consisted of a series of statistical tests to ascertain whether the difference between two means is or is not accidental—that is, whether the difference between two means is due to randomness or is indicative of some pattern. Using samples that would not necessarily qualify as random by today's strict standards, he compared the mean height of the general population to that of criminals and then to that of lunatics, the mean height of boys from upper middle class towns to that of boys from factory towns, the mean height of boys from public schools to that of boys from artisan towns, the mean height of men age 25–29 to that of men age 30–40, and the mean height of members of the Royal Society to that of members from the society "100 Sheffield." In each of these examples, Edgeworth determined that the observed differences in the average heights were not accidental but indicated differences that he deemed "significant" or "material." In fact, today we call a test to determine whether the difference between two means is significant, or merely a result of chance, a *test of significance*.

In another series of significance tests, Edgeworth compared
the proportion of female births to male births. These compari-
sons were performed by the age and occupation of the parent
and by the place, year, and season of birth. He found that
variations in the sex ratio due to the parent's age, occupation,
and year of birth were not at all surprising—indeed, his statistical tests indicated that they could be a result of chance. The
season of birth and place of birth, however, were factors that
seemed significant in terms of affecting variations in the sex
ratios.

Edgeworth also examined the mortality rates of English
males. He compared the mortality rate of those in different
occupations, the mortality rate of drinkers to the general population, and the mortality rate in England over several given time
periods. His tests confirmed that while differences in mortality
for drinkers and among occupations were significant, there was
no significant difference in mortality rates over various time
periods. Though Edgeworth took no great pains to assure that
his data samples were random, he claimed that in the case of
mortality data it has been "found in many cases that deaths at
certain ages fluctuate as I have said, and are distributed just like
damnatory lots at random taken from an urn."[8]

Edgeworth continued his statistical testing by examining the
mean differences in bankruptcies in different quarters of the

year (the tests demonstrated that the differences were significant), the mean differences in the quarterly amount of bills of exchange (the differences were not significant), the differences in the mean number of wasps entering or leaving their nest during different hours of the day (the differences were "insignificant"; today a statistician would use the term "not significant"), and the differences in the mean attendance at a London social club on different days of the week (the differences were "material"). In all of these samples, Edgeworth simply assumed his samples were random because there seemed to be no apparent bias in the way the data was gathered. Today we know that this assumption is not a sufficient precaution.

In a final example, Edgeworth devised a test to discriminate Virgilian style by differences in the mean number of dactyls in the hexameter. Using samples of lines in the *Aeneid,* he conducted tests to confirm passages that were consistent in style and to discriminate Virgilian from Ovidean hexameter. In a second paper published in 1885, Edgeworth indicated that samples of hexameter from the *Aeneid* were selected by using digits from a page of death rates. This may be the first mention of the use of a random number table.[9]

Edgeworth was perhaps more intent on elucidating statistical theory and demonstrating its wide applicability than on making great advancements in the theory itself. Indeed, one of his extraordinary contributions was to show his audience that statistical theory had applications to economic and social phenomena that went far beyond its previous uses.

In 1905 Walter F. R. Weldon, a zoologist at Cambridge University, delivered a series of lectures intended to demonstrate the laws of probability as applied to the Darwinian theory of natural selection in plant and animal hereditary factors. In these lectures he drew upon massive data from dice-throwing experiments that he had been conducting for many years with his wife.[10] In his opening experiment, Weldon demonstrated how accurately experimental results such as dice throws could be predicted by the laws of probability. For three independent sets of 4,096 tosses of twelve dice, Weldon compared the distribution of the number of dice with four or more spots showing with the results predicted by the laws of probability. The probability of four or more spots in a single toss is the same as the probability of three or fewer spots, 0.5.

Why 4,096 tosses? This is the smallest number of tosses of twelve dice for which we can expect that every possible outcome, no matter how extreme, will occur at least once. For example, in 4,096 tosses of twelve dice, we can expect (mathematically) one toss with all twelve dice showing four or more spots.

In this experiment, Weldon did not use 0.5 to compute the

probability of throwing a single die with four or more spots showing. He said that since no dice are symmetrical, he had based the probability of success in a single throw (0.509) on the frequencies he actually observed in the experiment.

In a second experiment, Weldon illustrated patterns in heredity between successive generations by using the correlation between the first and second throws of dice. In heredity, each offspring receives one half of its characteristics from each parent. If the set of characteristics that the child receives from the parent occurs at random, Weldon believed that he could model the patterns of heredity from one parent by a dice experiment. In a first throw of twelve dice—six painted red and six white—Weldon recorded the total number of dice with four or more spots showing. Next, leaving the six red dice on the table, he rolled the six white dice a second time. Once again the total number of dice showing four or more spots was recorded, including those on the six red dice that had remained on the table.

A pair of numbers (the number of dice showing four or more in the first roll and the number of dice showing four or more in the second roll) defined one trial in the experiment. The experiment recorded 4096 trials of two tosses of twelve dice, performed each time so that the six red dice left on the table constituted part of the second throw. Since the number of dice showing four or more in the second throw was *not* independent

of the first throw, Weldon felt this dice experiment illustrated
the correlation between inherited characteristics of an offspring
and one of its parents.

Weldon used these artificial data as a teaching tool, since it
was so easy for an audience to see how a pattern could arise from
even random data like dice throws. Figure 19 displays Weldon's
experimental results. The columns are labeled 0 through 12,
indicating outcomes possible on the first throw, and the rows are
labeled 12 through 0, the outcomes possible on the second
throw. The numbers in the cells of the table represent the total
counts of particular outcomes on the first and second throws.
We can clearly see the strong pattern in this data. When the
number of dice showing four or more spots on the first throw is
small, the number of dice showing four or more spots on the
second throw tends to be small; when the number of dice showing four or more spots on the first throw is large, the number of
dice showing four or more spots on the second throw tends to
be large. This *correlation* accounts for the pattern we see in the
table: the largest counts of trials appear to cluster from the lower
left corner of the table to the upper right corner.

In 1907 A. D. Darbishire performed an interesting series of
experiments in an "attempt to make the phenomenon of correlation clear to an audience previously unfamiliar with it."[11] Darbishire expanded on Weldon's experiment in order to illustrate

Number of dice showing four or more spots on the 1st roll

2nd roll	0	1	2	3	4	5	6	7	8	9	10	11	12	Totals:
12														0
11							1	1	5	1		1		9
10						2	6	28	27	19	2			84
9				1	2	11	43	76	57	54	15	4		263
8				6	18	49	116	138	118	59	25	5		534
7				12	47	109	208	213	118	71	23	1		802
6			9	29	77	199	244	198	121	32	3			912
5		3	12	51	119	181	200	129	69	18	3			785
4		2	16	55	100	117	91	46	19	3				449
3		2	14	28	53	43	34	17	1					192
2			7	12	13	18	4	1	1					56
1			2	4	1	2	1							10
0														0
Totals:	0	7	60	198	430	731	948	847	536	257	71	11	0	4,096

(Left axis label: Number of dice showing four or more spots on the 2nd roll)

FIGURE 19 Walter F. R. Weldon's 1905 dice-throwing experiments illustrating the correlation between successive rolls of twelve dice.

thirteen levels of correlation. Just as Weldon had done, Darbishire rolled twelve dice and recorded the number of dice showing four or more spots. Some of the dice were rerolled and, along with those left on the table, constituted the second throw, where once again the number of dice showing four or more spots was recorded. As in Weldon's experiment, the rolls were correlated, since the dice left on the table constituted part of both throws.

Darbishire explained that a number, called the *correlation coefficient,* measured the degree of connectedness between the first and second dice throws. Whereas Weldon's demonstration left six dice on the table from the first throw and six were rerolled, Darbishire performed thirteen different demonstrations of the twelve-dice rolls. In each of the thirteen experiments, after rolling all twelve dice, Darbishire left a different number of dice from the first throw on the table and rolled the remainder. Before beginning an experiment, he stained the dice that were to remain on the table a different color. The first experiment involved leaving no dice on the table (throwing all twelve again), the second involved leaving one die on the table (eleven were rethrown), and so on, so that in the thirteenth experiment all twelve were left on the table (and zero were thrown in the second throw).

When no dice are left on the table and all twelve are rethrown, there is no connection between the first and second throws, and the correlation coefficient is 0. If six of the twelve dice are left from the first throw to constitute part of the second throw (Weldon's experiment), the correlation coefficient is 0.5. If all twelve of the dice in the first throw constitute the second throw, the throws are perfectly correlated, and the correlation coefficient is 1. Each of Darbishire's thirteen experiments involved 500 recordings of the number of dice showing four or more spots on the first and second rolls of the dice. His results were summarized in thirteen tables of first throws versus second throws, similar to Weldon's table, which demonstrate thirteen degrees of correlation—ranging from no correlation to perfect correlation.

Chi-squares and t-distributions

One of the most influential of Britain's early statisticians was Karl Pearson, who was a colleague of Edgeworth and a disciple of Galton. Pearson had constructed a device similar to Galton's quincunx as early as 1895, but his work does not seem to contain any random sampling experiments prior to 1900. In that year he published his influential chi-square goodness-of-fit test—a statistical test whereby a set of data can be examined to confirm how well its distribution actually fits a distribution one believes it to have. Pearson's test allowed any set of data to be compared with a hypothesized distribution to determine its goodness of fit to that distribution. He noted that if the deviation from the hypothesized distribution is small, "it can be reasonably supposed to have arisen from Random Sampling."[12]

To illustrate his new chi-square method, Pearson demonstrated that the runs at roulette occurring at Monte Carlo in July of 1892 did not resemble a distribution of runs that would occur purely by chance. Pearson said that the odds were at least 1000 million to one against such an outcome occurring as the random result of a true roulette. In other words, most likely the game was rigged.

In another demonstration, Pearson compared 26,306 tosses of twelve dice to the distribution expected if the dice were fair and the results truly random. The data being tested were the number of dice showing a five or a six in each toss. Pearson concluded that the probability that the results were produced by chance was extremely remote: the data showed more fives and sixes than one ought to see in a chance experiment. Pearson had obtained his dice-throwing data from Walter F. R. Weldon. If Weldon's data were in fact nonrandom, it might be because there is a tendency in all observations, scientific and otherwise, to see what one is looking for (in this case fives and sixes) more often than it was actually there.

Weldon, who must have felt that this conclusion reflected

badly on the great care and precision he had taken in his dice-throwing experiments, suggested that Pearson ought not to use the theoretical probability of a five or six showing on one roll of a die (probability = $\frac{1}{3}$), but ought to use rather the relative frequency actually observed in the experiment, 0.3377, since the dice may not have been symmetrical. As a concession to Weldon, Pearson performed his goodness-of-fit test using 0.3377 and concluded that Weldon's data *were* consistent with a chance experiment in which the chance of a five or six showing was 0.3377 but not one in which all sides of the die were equally likely. An interesting footnote to this episode is that Pearson inadvertently erred somewhat in Weldon's favor in performing this second test.

As a further illustration of his new goodness-of-fit test, Pearson tested several sets of data that were presumed to exhibit a normal distribution. Most of these data failed the test. One set of data that he developed himself—errors in bisecting 500 lengths by sight—did conform to the normal law. He concluded "that the normal curve possesses no special fitness for describing errors or deviations such as arise either in observing practice or in nature."[13]

In his highly influential 1900 article, Pearson severely criticized earlier scientists for proceeding on the assumption that they were dealing with normally distributed data without having obtained scientific verification of that assumption. According to Pearson, the theory of errors, or normal distribution, was *not* a universal law of nature. In fact, very few samples assumed to be normal turned out to be normally distributed when his good-ness-of-fit test was applied. Even dice throws, as he demonstrated, may not be random.

The British statisticians surrounding Pearson at University College must have taken this message to heart. They were no longer willing simply to use samples assumed to be normal or assumed to be random but began creating their own random samples.

The best-known early demonstration of a random sampling experiment was performed by William Sealy Gosset, a research chemist working for the Arthur Guinness Son and Company Ltd. in Dublin.[14] Gosset was studying the relationship between the quality of Guinness beer and various factors in the beer's production. The brewery was continually experimenting with soil conditions and grain variety that might produce improvements in crop yield, and Gosset was intent on bringing all the benefits of statistics to the brewery's agricultural experiments.

Since most of the land was given over to actual production of hops, experiments were necessarily small-scale. As a result, Gosset's samples consisted of only a few pieces of data—sometimes limited to eight or ten observations. Up to that time, most of the results in statistics were applicable to large samples of data but were not valid for small samples unless a good deal was known about the population from which the samples were drawn. Gosset found that he needed statistics that were valid when the population was not well understood and only small samples could be obtained. Gosset felt that others like himself, who were performing chemical, biological, agricultural, and other large-scale experiments under the constraints of small samples, might well find his results useful. Aware, however, that Guinness would not allow its employees to publish, lest company secrets be divulged to the competition, Gosset published under the pseudonym Student. Consequently, his statistical discovery, the t-distribution, has become known as Student's t.

Gosset's best-known paper, written in 1908, described Student's t, which is defined as the *distribution of means in small samples*.[15] Since the existing methods of analyzing means (namely, the normal curve and the Central Limit Theorem) pertained only to large samples, one aim of his paper was to define "large." A second goal was to produce a statistical method that could be used by investigators who must rely on small samples from normal populations about which little was known.

In order to examine the distribution of means of small random samples, Gosset created 750 random samples, with each sample containing four observations. Obviously Gosset could not use any of Guinness's secret agricultural data, and the data he chose for this experiment are rather unusual: the height and finger length of criminals![16] His samples were created by writing the height and left middle finger length for each of 3000 criminals on pieces of cardboard, thoroughly shuffling the pieces of cardboard, and drawing 750 samples of four observations each. From this experiment involving the finger length of criminals, Gosset made an important new advance in applying statistics to areas where we have limited knowledge of the population under investigation.

When applying the Central Limit Theorem to means of even large samples, a value needed in the calculations is the *standard deviation,* which measures the variability of the population under study. Gosset discovered the distribution for means of samples when the standard deviation was derived from the sample itself, rather than from the population as a whole. Hence, Student's t-distribution could be used when the standard deviation in the total population was unknown, even when samples were very small.

Figure 20 presents a graph of the Student's t versus the normal curve for sample means taken from samples of size 10. In both cases, probabilities are measured by computing the areas under the curves. Notice the similarity between the normal curve and Student's t. The horizontal axis is measured in devia-

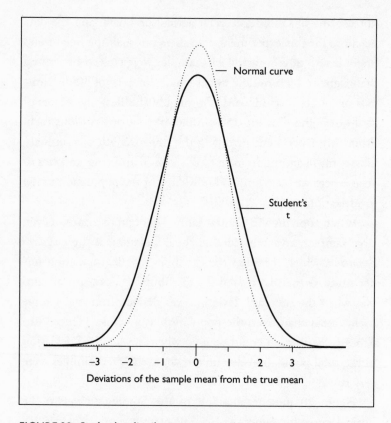

FIGURE 20 Student's t-distribution, compared with the normal curve.

tions from the mean. If available, this unit is computed using the size of the population standard deviation, and sample means will follow the normal distribution. If the population standard deviation is not available, the standard deviation must be computed from the sample itself. Because small samples are highly variable, when that statistic is used in statistical tests that assume a normal distribution, results may be misleading, and conclusions may be exaggerated and unjustified.

Gosset found that the means of small samples would more nearly follow the pattern of his t-distribution than the normal distribution. To the untrained eye, these curves may appear quite similar, and indeed the larger the sample, the closer the resemblance between the two curves will be. When sample sizes are small, however, the differences prove critical to arriving at accurate statistical conclusions.

In another article in the same journal, Gosset went on to investigate correlation in small samples using the height and finger length data of criminals. Gosset states that he wishes "to throw some light by empirical methods" on the difficulty of using a correlation coefficient derived from small samples. His results are inconclusive, and he states that he presents the results for the interested mathematician to eventually solve the problem.[17]

Gosset was remarkably influential, as can be seen in the work

of J. W. Bispham (at that time Captain J. W. Bispham), to take
just one example. In 1914 Bispham began an investigation into
correlation in small samples but put it aside because of World
War I. He eventually reported his findings in 1920 and 1923,
stressing that his experimental investigation with small samples
was undertaken to examine correlation in indices such as those
used in studying mortality rates. His conclusions indicate that
no serious differences result from using an alternative statistic in
place of the correlation coefficient in even small samples.

More interesting than his conclusions is the inordinate
amount of work that went into creating the random samples for
his experiments. Bispham obtained data by having school children randomly draw numbered bone discs, or "counters," from
a container. In the first study, 1000 different times children
made three different draws of 30 counters. In the second study,
10 counters were drawn and summed 3000 times, 30 counters
were drawn and summed 6000 times, and 60 counters were
drawn and summed 3000 times.

Bispham noted, "The preliminary arithmetical work has
been extremely laborious. It has involved the handling of individual 'counters,' of something of the order of half a million
times and the subsequent summing in groups of ten, thirty or
sixty."[18] Creating genuine random samples for experimentation

was apparently becoming quite burdensome. A new method of
generating random samples clearly was needed.

By the twentieth century, statistical experimentation had gained
a certain acceptance and status, due in large part to the more
precise methods of random sampling and the creation and use
of artificial random data. Gosset's work was well known and
respected, and his random sampling experiment was highly
emulated. A spirit of empirical experimentation and demonstration was emerging, though the task of creating genuine random
samples with dice, counters, and slips of paper was becoming
quite onerous.

Moreover, by the early decades of the twentieth century, the
notion of randomness itself was changing. Random selection
could no longer mean haphazard selection. Though there was
no single accepted definition, randomness (or the lack of it)
could now be measured.

Wanted: Random Numbers

8

In the 1920s, Leonard H. C. Tippett, in the course of demonstrating a new statistical idea, needed a large set of random data. At first he attempted to generate 5000 random numbers by mixing and drawing small numbered cards from a bag. This method proved unwieldy, and the mixing of the cards was not sufficient to ensure randomness.

Next, Tippett developed data for this demonstration by obtaining 40,000 digits "taken at random" from the areas of census parishes. In his 1925 paper describing his work, Tippett stated that these digits could be used to construct random samples from any population, and expressed his hope that someone would publish them.[1]

As Alfred Bork has pointed out, "A rational nineteenth-century man would have thought it the height of folly to produce a book containing only random numbers." Nevertheless, in 1927

```
0153 7051 2272 1359 3328 0014 6773 1278
6282 1805 5034 6723 3835 6978 7084 3992
7542 2529 0311 2979 0095 2647 8299 5163
7757 5430 4866 6497 4138 8144 0294 2906
4151 3879 3062 7604 8137 4575 2245 6309

0764 9357 2633 8605 2064 0736 3046 0612
4691 3656 9675 0286 6825 7823 5778 2680
4057 0762 6469 2735 5082 3852 7457 5729
5484 0770 7222 4912 0062 0609 9291 4056
0125 9592 3729 7858 5153 7200 1308 9638

5587 2698 2748 0458 0122 4721 3963 2916
7963 1937 6002 4490 5494 2817 6818 7120
8894 0546 6771 8401 1359 9935 8594 7513
9090 2972 0932 3907 6077 7374 0992 8951
7986 0132 8683 8568 2374 4215 3574 4177
```

FIGURE 21   A portion of Leonard H. C. Tippett's 1927 table of random digits. (Reproduced by permission of Cambridge University Press.)

Cambridge University Press did indeed publish a table of 41,600 digits that had been randomly arranged by Tippett (see Figure 21).[2] As Tippett's mentor, Karl Pearson, stated in the foreword, "A very large amount of labour has been spent in recent years in testing various statistical theories by aid of artificial random samples." Rejecting the effectiveness of cards, tickets, balls, and dice, Pearson contended that statistical experimenters "who have had to deal with the problems of random sampling" might benefit by "a single system of numbers." The table, which was already being used extensively in the Department of Applied Statistics at University College, London, where Tippett was studying, became known as Tippett's Random Sampling Numbers. It is the first published table of random digits.

A mere ten years after its publication, Tippett's table of over 40,000 Random Sampling Numbers was deemed inadequate for very large sampling experiments. In 1938 the mathematicians R. A. Fisher and F. Yates published 15,000 additional random digits, selected from the 15th through the 19th decimal places of logarithm expansions. The digits were obtained through a procedure involving two decks of cards. In 1939 M. G. Kendall and B. Babington-Smith published a table of 100,000 digits that were ordered randomly by a machine constructed from a rotating disc. The disc was divided into ten sectors, and as the disc rotated, one of the ten sectors was momentarily illuminated by a flashing neon lamp. In 1942 J. G. Peatman and R. Shafer published 1600 random digits obtained from Selective Service lotteries.[3]

In 1949 the Interstate Commerce Commission published a table of 105,000 random digits, generated by a process called *compound randomization*. Writing for the ICC, H. Burke Horton stated that previously generated digits like Tippett's had

suffered the same biases as the single-stage mechanical or electronic devices that generated them. In an attempt to eliminate
this bias, Horton demonstrated that random digits could be
produced from sums of other random digits and that the com-
pounding of the randomization process created a sequence with
less bias than the original sequence.[4]

For example, take two random sequences (of zeroes and
ones) of the same length where each digit is equally likely, say
0111101100 . . . 1 and 1011111100 . . . 0. We will add them
together according to the following rules: 0 plus 0 equals 0; 1
plus 0 equals 1; 0 plus 1 equals 1; and 1 plus 1 equals 0. This
addition yields a new sequence of zeroes and ones. In each
position of the new sequence, a 0 or 1 is equally likely (out
of four possible sums, two yield 0 and two yield 1). Horton
claimed that this new sequence, 110001000 . . . 1, would be
"more random" than either of the original two.

In 1955 the RAND Corporation published a document entitled *A Million Random Digits with 100,000 Normal Deviates.*
This document was developed by rerandomizing a table of digits
generated by the random frequency pulses of an electronic roulette wheel. RAND stated that the purpose of producing such
large tables was to meet the growing need for random numbers
in solving problems by experimental probability procedures.[5]

It had become clear that the tables of random digits produced so far were not nearly large enough to satisfy the vast number of applications that were evolving. RAND's huge table of random digits had forecast the need for bigger and bigger tables as new methods of modeling and simulation developed. Sources of random digits other than published tables were also considered, sources which would be readily available to the scientist as demand increased. For example, a physical process like the electronic impulses of a computer or the decimal expansion of certain irrational numbers might produce an endless source of random digits.

With the availability of digital computers, more and more random numbers were required, not just in sampling, which had become an integral part of experimental design, but also in models that predicted complex trends. Probability theory was being used to model elements of uncertainty in economic forecasting, decision theory, inventory and queuing theory, and biological, sociological, and physical theory. In addition, experimental probabilistic methods were also being devised to solve difficult deterministic (nonprobabilistic) problems using a procedure known as the Monte Carlo method.

The Monte Carlo method provides approximate solutions to problems considered far too difficult to solve directly. It solves a probabilistic problem similar to the nonprobabilistic one through experimental trials. The Monte Carlo method goes back to Buffon, who in his 1777 *Essai* had described an experiment in geometric probability that became known as "Buffon's needle problem." Having laid off two parallel lines on a plane surface like a table, Buffon's problem asked for the probability that a needle tossed randomly upon the surface would cross one of the lines. In 1820 Laplace renewed interest in this problem—going on to suggest that with a large number of tosses, one could estimate the value of 2π, using the theoretical probability and the experimental results.[6]

There is no indication that either Buffon or Laplace actually tried the experiment, but each is given credit for the suggestion of what we would today call a Monte Carlo experiment—a method of estimating a constant quantity like π through probabilistic procedures and experimental trials involving random results.[7] The first evidence of an actual investigation of this type occurred when Richard Wolf performed Buffon's experiment in 1850 with 5000 tosses of the needle. Subsequent experiments followed: Ambrose Smith in 1855 with 3204 tosses, Augustus De Morgan's student, circa 1860, with 600 tosses, Captain Fox in 1864 with 500, 530, and 590 tosses, Mario Lazzerini in 1901 with 3408 tosses (using a mechanized device to toss the needle), and Reina in 1925 with 2520 tosses. It has been pointed out that their estimates of π are "absurdly accurate" considering the number of trials they performed. Between 1933 and 1937, A. L.

138   Clark had students perform 500,000 trials of an experiment to estimate π by dropping balls through a circular opening. He was convinced that the earlier experimenters conveniently decided when to stop their experiments, depending on how good their estimate was.[8]

The Monte Carlo method came into widespread use only with the advent of the computer, since the computer can easily simulate a large number of experimental trials that have random outcomes. While working at Los Alamos on the hydrogen bomb, the mathematician and computer pioneer John von Neumann, along with Stanislaw Ulam, originally implemented the Monte Carlo method to determine statistically the many random components at each stage of the fission process.[9]

Suppose we need to solve a difficult mathematics problem, such as finding the area under an irregular curve. Sometimes this kind of problem can be solved by using calculus, but if the problem proves intractable to higher mathematics, it can be solved approximately with other techniques, including the Monte Carlo method. A peculiar aspect of this method of problem solution is that, while the problem requires an exact solution—the area is fixed, there is nothing probabilistic about it— the Monte Carlo method relies on *randomness*.

To find the area under an irregular curve, we surround the area with a rectangle and instruct the computer to generate a

large number of points at random positions within the rectangle
(see Figure 22). We then count the number of points lying
inside the boundary of the irregular curve and divide it by
the total number of points generated inside the rectangle. This
gives us the proportion of points that lie in our region of interest. Since computing the total rectangular area is easy (base ×
height), multiplying the proportion by the total area gives the
approximate area under the irregular curve. For increased reliability in the approximation, we generate a larger number of
random points.

As larger and larger tables of random digits were needed in
computer models to solve nonprobabilistic problems, not to
mention the wide range of probabilistic applications, storage of
the random number tables in computers began to consume far
too much memory. The use of a preprogrammed formula that
allowed the computer to generate a random digit at the time it
was needed in a calculation seemed to be the ideal solution.

Conceptually, the idea of an arithmetically determined random digit was both desirable and undesirable. The repeatability
of digit sequences allowed some control over modeling—the
modeler could vary specific parameters while replicating the rest
of the simulation exactly, using the same sequence of random
numbers. In other words, if we know how to get the identical
random sequence each time the simulation is performed, we can

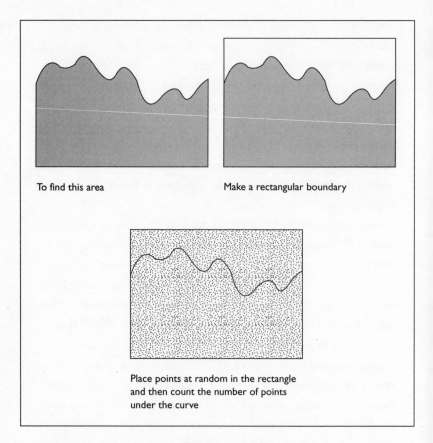

To find this area

Make a rectangular boundary

Place points at random in the rectangle
and then count the number of points
under the curve

FIGURE 22 Using the Monte Carlo method to find the area under an irregular curve.

determine the effect of changing certain parameters, without **141**
worrying about interference due to changes in our random sequence.[10] The undesirable aspect of computer-generated digits
resided in the deterministic nature of the process.

Within any sequence generated by the computer through a
programmed algorithm or formula, the next digit is a completely deterministic choice, not random in the sense that a dice
throw, a spinning disc, an electronic pulse, or even the infinite
digits of the mysterious π are random. The very notion that
a deterministic formula could generate a random sequence
seemed like a contradiction. Even von Neumann admitted this,
with his remark: "Anyone who considers arithmetical methods
of producing random digits is, of course, in a state of sin."
Nonetheless, this was precisely his sin when in 1946 he first
suggested using the arithmetic operations of a computer to produce random digits.[11]

Then in 1951 von Neumann proposed an arithmetical procedure to generate random numbers: an algorithm, called the
middle-square method, which generates groups of n random
digits by starting with a number n digits long, squaring it, and
taking the middle n or $n+1$ digits for the first group, squaring
this number and taking the middle n or $n+1$ digits for the
second group, and so on.[12] For example, the 3-digit number
123, when squared, yields 15,129. The middle three digits of

15,129 are 512. Squaring 512 yields the number 262,144. The 4 middle digits, 6214, when squared, yields 38,613,796, whose 4 middle digits are 6137, and so on. Our "random" sequence thus begins with 512|6214|6137.

Unfortunately, this method has been found to be a poor source of random numbers, flawed by imprudent choices of the original n digits. For example, starting with the 4-digit number 3792 and squaring it produces the number 14,379,264, with middle digits 3792—the same 4 digits we began with![13] And so the search began for algorithms that could be computer-coded into programs generating random digits. The digits were called pseudo-random numbers, and the programmed formulas, or algorithms, were called pseudo-random number generators—for how could truly random numbers be generated through a formula and a machine?

Most generators used a recursive formula, where each new digit generated was in some way based on the digit generated prior to it. Given an initial number (called a seed), the formula created a sequence of digits in apparently random order. Eventually, the sequence of digits started over, and the list began to repeat itself, or cycle. The length of the cycle, or the number of digits in the list that began to repeat, was called the period. Of course the longer the period the better. (If we need lots of

random digits, we don't want the digits in the same order over   **143**
and over again.)

Today, primarily three types of generators are in use: (1)
congruential generators, which are based on modular arithmetic, or remainders after division, (2) generators which use the
binary (bit) structure of computer-stored information, and (3)
generators based on number theory.[14]

Congruential generators use modular arithmetic, or the remainder after division, as the next digit in the sequence. For
example, a number mod 7 is replaced by the number's remainder after dividing the number by 7. 15 mod 7 is replaced by 1
since when 15 is divided by 7 (called the modulus), the remainder is 1; 18 mod 7 is replaced by 4 since 4 is the remainder
when 18 is divided by 7; 30 mod 7 is replaced by 2; and so on.
Generators using this method were first proposed by D. H.
Lehmer in 1949. If Z represents one of the digits 0 through 9,
then $Z_1|Z_2|Z_3|$ . . . $|Z_n$ represents an n-digit sequence of num-
bers. In order to produce the random sequence of digits
$Z_1|Z_2|Z_3|$ . . . $|Z_n$, the congruential generator uses modular arithmetic and the formula, $Z_{n+1} = aZ_n$ mod m, where the modulus
m determines the maximum length of the period. In reality, m
must be a very large prime number to get a useful generator with
a long period. Lehmer suggested the number 2,147,483,647 as

modulus. The multiplier *a* (for a given *m*) influences the actual length of the period, the apparent randomness of the sequence, and the ease of implementation on a computer.[15]

Let's see what digit sequence the Lehmer generator would produce with 7 as the modulus and 3 as the multiplier (see Figure 23). Since, for illustration purposes, we are using a small number as the modulus, the period will be short; it can be no longer than 7. With this modulus and multiplier, our generator consists of the formula $Z_{n+1} = 3Z_n$ mod 7. If we choose 127 as the initial seed, Z_0, the first random digit, Z_1, is produced by $Z_1 = 3(127)$ mod 7, or 381 mod 7. When 381 is divided by 7, the remainder is 3. Therefore, Z_1 is 3. As the seed, Z_0, was used to obtain Z_1, Z_1 is used to obtain the next digit, Z_2. $Z_2 = 3(Z_1)$ mod 7 = 3(3) mod 7 = 9 mod 7, and so Z_2 equals 2 (since 9 divided by 7 gives a remainder of 2). Because each new digit is based on the digit generated previous to it, this type of formula is called recursive. Our formula produces the sequence 3|2|6|4|5|1, which will repeat, cycling forever.

A new class of number theoretic generators has recently been developed by George Marsaglia and Arif Zaman. Emphasizing the simplicity of the computer arithmetic and modest storage requirements of their new generators, these researchers believe their generating procedures merit serious consideration for Monte Carlo work.[16] Called add-with-carry and subtract-with-

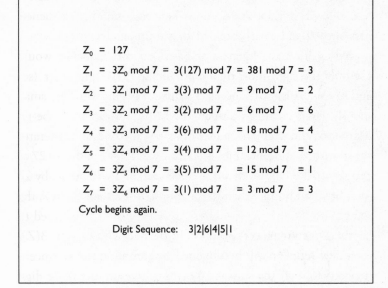

$$Z_0 = 127$$

$$Z_1 = 3Z_0 \bmod 7 = 3(127) \bmod 7 = 381 \bmod 7 = 3$$

$$Z_2 = 3Z_1 \bmod 7 = 3(3) \bmod 7 \quad = 9 \bmod 7 \quad = 2$$

$$Z_3 = 3Z_2 \bmod 7 = 3(2) \bmod 7 \quad = 6 \bmod 7 \quad = 6$$

$$Z_4 = 3Z_3 \bmod 7 = 3(6) \bmod 7 \quad = 18 \bmod 7 \quad = 4$$

$$Z_5 = 3Z_4 \bmod 7 = 3(4) \bmod 7 \quad = 12 \bmod 7 \quad = 5$$

$$Z_6 = 3Z_5 \bmod 7 = 3(5) \bmod 7 \quad = 15 \bmod 7 \quad = 1$$

$$Z_7 = 3Z_6 \bmod 7 = 3(1) \bmod 7 \quad = 3 \bmod 7 \quad = 3$$

Cycle begins again.

Digit Sequence: 3|2|6|4|5|1

FIGURE 23 Using a congruential generator to generate the random digit sequence 326451.

borrow generators, their technique relies on the Fibonacci sequence and the so-called lagged-Fibonacci sequence, both named after Leonardo Fibonacci, a twelfth-century Italian mathematician who, among other things, advocated the adoption of Arabic numbers.

The Fibonacci sequence consists of the numbers 0, 1, 1, 2, 3, 5, 8, 13, 21, 34, 55, and so on, where each number except the first two (0 and 1) is obtained by summing the two numbers preceding it. A digit-generator based on this sequence would use only the right-most digits, producing 0|1|1|2|3|5|8|3|1|4|5, and so on.[17] A lagged-Fibonacci sequence begins with two enormously large starting numbers, or seeds, instead of 0 and 1. Marsaglia and Zaman's add-with-carry method is similar to these with a modification called a "carry bit" and is simple enough to be employed on a calculator or by hand. As with long-hand addition, whenever a sum is more than 9, we carry the 1.

As in the Fibonacci sequence, in the add-with-carry method each new number will be obtained by summing the two digits previous to it. If the sum is 10 or more, we use the right-most digit only and carry the 1 (to be used in obtaining the next digit). So, each new digit in the sequence is formed from the sum of the two previous digits plus a 1 if it was carried. For instance, beginning with the two initial seeds 0 and 1 (see Figure 24), we obtain the same beginning sequence as the Fibonacci sequence 0, 1, 1, 2, 3, 5, 8, until we reach 13. Here, the 3 is used and the 1 is carried. The next number in the sequence is obtained by summing the previous two, 8 and 3, and the 1 that was carried. 8 + 3 + 1 is 12, so the 2 is used and 1 is carried. 3

```
        Sum two previous digits + carry?   =   sum ,   digit

                        0 + 1               =    1 ,     1

                        1 + 1               =    2 ,     2

                        1 + 2               =    3 ,     3

                        2 + 3               =    5 ,     5

                        3 + 5               =    8 ,     8

                        5 + 8               =   13 ,     3*

                        8 + 3      +  1     =   12 ,     2*

                        3 + 2      +  1     =    6 ,     6

                        2 + 6               =    8 ,     8

                        6 + 8               =   14 ,     4*

                        8 + 4      +  1     =   13 ,     3*

                        4 + 3      +  1     =    8 ,     8

                        3 + 8               =   11 ,     1*

                        8 + 1      +  1     =   10 ,     0*

                        And so on.
        The digit sequence thus far is 1|2|3|5|8|3|2|6|8|4|3|8|1|0.
                                                        * carry the 1
```

FIGURE 24 Using the add-with-carry generator to generate the random digit
sequence 12358326843810.

148 + 2 + 1 is 6 which becomes the next digit and (since 6 is less than 10) nothing is carried.

Since the purpose of using a computer program to generate the digits of a random sequence is to avoid having to store a large table of digits, from the point of view of a computer scientist, the shorter the program the better. Still the search continues—for better generators, for faster generators, and for generators with longer periods. As Donald Knuth put it, "Random numbers should not be generated with a method chosen at random."[18]

New Uses for Random Digits

While computer power has made more random numbers possible, it has also created an insatiable appetite for them. For example, computerized data-based simulation methods for statistical inference, called resampling techniques, have generated a huge demand for random digits. One such resampling procedure, known as bootstrapping, was introduced in 1979 by Bradley Efron. Refined and discussed by theoretical statisticians for fifteen years, bootstrapping may well replace the traditional method of inference which compares a single random sample to hypothesized theoretical samples. Bootstrapping does not rely on knowledge of the theoretical properties of statistical measures but instead relies on the power of the high-speed digital computer.

In traditional inferential statistics, after a statistical measure is estimated from a sample drawn randomly, its reliability is assessed based on the theoretical distribution of the statistic—a distribution such as the normal distribution or Student's t-distribution. In the bootstrap procedure, after a statistical measure is estimated from a sample drawn randomly, its reliability is assessed based on thousands of real samples which are generated by the computer. The new samples, called bootstrap samples, are drawn from the one original sample which has been mixed, and as each item is selected it is replaced and eligible for selection again. The statistical measure being estimated is computed for each bootstrap sample, and these serve in the same way that the theoretical distribution does in traditional inferential statistics to assess the reliability of the original estimate.[19]

New procedures for keeping secrets, which have important applications in national security and in electronic information systems, require many random components. These methods of cryptography, called public key, require random selection of 100- or 200-digit prime numbers; the unpredictability of the numbers chosen along with the fact that some arithmetic operations are very difficult to reverse can ensure that the code would be extremely difficult to break in a reasonable amount of time. A new method of "proof," called the zero-knowledge or interactive proof, which can convince a second party that information is held without revealing any of the information, requires random selection of a series of questions by the second party. The ability of the "proof" to convince the second party is based on the probability that information requested at random would not be able to be provided unless the entire proof was known.[20]

Chaos theory, the science which predicts that the future state of most systems is unpredictable due to even small initial uncertainties, holds new meaning for the notion of randomness, and simulating these systems requires huge numbers of random digits. It has been shown that with even small deterministic systems, initial observational error and tiny disturbances grow exponentially and create enormous problems with predictability in the long run.[21]

Virtually since the idea arose to maintain a storehouse of random digits, scientists have been coming up with more and better ways of using them. Today, random numbers are needed for economic models (predicting when the bond market will decline and by how much) and traffic models (predicting when a car will reach an intersection or when a car will run a red light).

Computer models that simulate molecular movement or the   **151**
behavior of galaxies require vast quantities of random numbers.

High-speed computers have produced both a need for random digits that could be available immediately and a method by which those digits could be produced via algorithmic generators. These generators, if reliable, promise that the digits have been thoroughly shuffled and each digit is equally likely to be drawn. Still, the notion that a machine and a programmed formula can create randomness makes some mathematicians very uncomfortable.

Randomness as Uncertainty

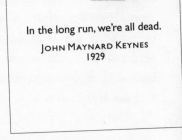

The break at the beginning of a pocket billiards game sends colored balls rolling, ricocheting off one another, and bouncing off cushions until they all finally come to rest in what would appear to be a random pattern. Yet physical law can predict the path of one ball hitting another, hitting a cushion, rolling on an inclined table, rolling over a bare spot on the table, rolling in humid weather, or rolling over dust on the table. Given the particular initial conditions and all the facts involved, these paths and final positions can be determined mathematically.

Does this mean that the distribution of the billiard balls at rest is not random? Can randomness result from nonrandom situations? Is randomness merely the human inability to recognize a pattern that may in fact exist? Or is randomness a function of our inability, at any point, to predict the result?

Cicero notes that the main feature of luck or chance is

uncertainty. "For we do not apply the words 'chance,' 'luck,' 'accident,' or 'casualty' except to an event which has so occurred or happened that it either might not have occurred at all, or might have occurred in any other way. For, if a thing that is going to happen, may happen in one way or another, indifferently, chance is predominant; but things that happen by chance cannot be certain."[1] Cicero characterizes chance events as events which occurred or will occur in ways that are uncertain— events that may happen or not happen, or may happen in some other way.

This quite "modern" view was not typical prior to the twentieth century. The more widespread belief was that what we call chance is merely ignorance of initial conditions. The popularity of this deterministic interpretation is not surprising if we consider that the study of random variation originated with the theory of errors: that differences in observation or measurement of a single fixed quantity, such as the distance to a star or a specific distance on the earth, were merely errors. In these cases, scientists used the average of the measurements as the "true" distance, and variation among individual measurements was considered to be the result of inadequacies on the part of the measurers or the measuring equipment.

When this view was applied to games of chance, variation among trials was considered to be an indication of the limitations of human knowledge. When applied to social phenomena, the notion of random variation as ignorance or error must have been more difficult to justify, but researchers still considered the average measurement as the "ideal" and variation as deviations from that ideal.

In his 1739 work *A Treatise of Human Understanding*, the Scottish philosopher David Hume discussed chance from a different perspective—its effect upon the mind. He said that chance leaves the mind "in its naive situation of indifference." That is, a set of equally likely outcomes produces a mental state of indifference among alternatives, and there is no reason to prefer one outcome over another.[2] The belief that chance represents either insufficient knowledge or indifference is sometimes referred to as the subjective definition of randomness: according to this view, randomness exists only in the minds of individuals, not in the objective world.

John Stuart Mill's *A System of Logic*, published in 1843, took issue with a theory of probability based on a balance of ignorance or indifference. Mill found it strange indeed that ignorance and subjectivity should be the basis of a "science." Mill thought that the probability of an event should be based rather on our knowledge and experience of frequencies. He insisted, "To be able to pronounce two events equally probable, it is not enough that we should know that one or the other must happen, and should have no ground for conjecturing which. Experience must have shown that the two events are of equally frequent occurrence."[3]

John Venn, the English logician after whom the Venn diagram (used to pictorially represent sets) is named, admired Mill's work, adding that his own philosophy was similar, "taking cognisance of laws of things and not of the laws of our own minds in thinking about things." He extended the frequency view from past experience to future (long-run) frequencies as he advanced the hypothesis that a random selection process requires that each member of the universe shall, in the long run, be selected equally often. Random sequences are often ones whose elements are equally likely, and Venn asserted that an arrangement ought to be judged random or not by "what we have reason to conclude would be observed if we could continue our observation much longer."[4]

Along the same lines, in 1896 Charles Sanders Peirce, the American physicist and philosopher, offered a definition of a random sample as one "taken according to a precept or method which, being applied over and over again indefinitely, would in the long run result in the drawing of any one set of instances as often as any other set of the same number."[5]

Some philosophers, such as John Maynard Keynes, rejected the frequency theory outright, claiming that a definition of random selection ought to be more useful, more practical. Keynes objected that "in the long run" required complete knowledge.[6]

Reflecting this controversy, the 1902 edition of the *Encyclopaedia Britannica* presented both the subjective and the frequency interpretation of an equally likely probability distribution.[7] A simple toss of a fair coin illustrates the differences between the two. A physical interpretation of the reason that the probability of a head is $\frac{1}{2}$ is that there are two sides, the coin is balanced, and physical laws enable the coin to fall equally well on either side. But this begs the question; one might well ask how do we *know* that it can fall equally well on either side? According to the subjective view, since we don't know which of the two sides will land face up and we are equally indifferent between the possibilities, before we throw the coin the likelihood of each outcome is equal, $\frac{1}{2}$. The frequentist, who is uncomfortable basing a mathematical probability on ignorance or frame of mind, will say that the probability is $\frac{1}{2}$ because experience has shown that over a long period of time the head as well as the tail has come up about one-half of the time.

In the 1888 edition of *Logic of Chance*, Venn attempted to create a visual illustration of randomness by building a randomly generated graph. Each step of his "random path" was taken by allowing movement chosen at random in any of eight directions (the eight points of the compass; see Figure 25, left).

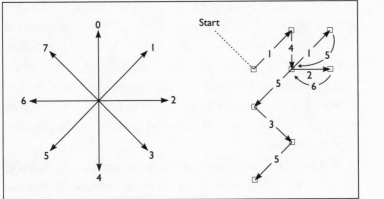

FIGURE 25 After assigning a direction of movement to the digits 0 through 7, John Venn determined each step along this random path by using the initial digits in the decimal expansion of π.

The direction of movement at each step along the path was determined by using the digits in the decimal expansion of π—a sequence of digits which Venn believed to be random. The first 707 decimal digits of π were used in the order in which they occurred—the digit 0 generated movement to the north; 1, to the northeast; 2, to the east; and the digits 3–7 continued in a clockwise direction around the points of the compass.

We can easily see how he began his path by looking at the first few digits (Figure 25, right). For example, if we take π =

3.1415926535 . . . , we would follow Venn's direction of movement compass for the digits 141526535. Venn discarded the
initial 3 and all 8s and 9s. Convinced of the path's patternless-
ness by visual inspection, Venn concluded that the graph of the
path generated by those digits would enable the reader to comprehend the idea of random arrangement (see Figure 26). Venn
indicated that he had drawn similar paths using digits from
logarithm tables.[8]

It is unknown whether the digits in the decimal expansion of
numbers like π, e, and √2̄ are arranged randomly, but the
attempt to verify or disprove this idea continues today. In their
belief that the decimal expansion of certain irrational numbers
might produce an endless source of random digits, others have
followed Venn's lead. In 1950 N. C. Metropolis, G. Reitwiesner,
and J. von Neumann published the results of a statistical test for
randomness on 2000 digits of π and 2500 digits of e. Their tests
concluded that while π failed to deviate from randomness, e
revealed serious deviations.[9]

In the same year, C. Eisenhart and L. S. Deming came to
similar conclusions, having applied tests to 2000 digits of π and
e. In 1951 Horace S. Uhler tested 1545 digits of √2̄ and 1/√2̄,
and his results indicated no departure from a purely random
group of digits. In 1955 Robert E. Greenwood tested 2035
digits of π and 2486 digits of e, concluding that his test showed

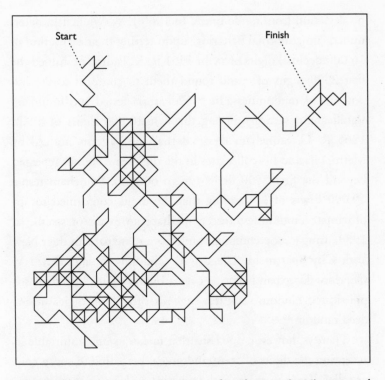

Start

Finish

FIGURE 26 Venn's visual representation of randomness (1888), generated from the first 707 decimal digits of π.

no deviation from randomness. In 1960 J. W. Wrench Jr. determined "no abnormal behavior" upon testing the distribution of 16,000 decimal digits of π. In 1961 R. K. Pathria examined the first 2500 digits of e and found them to conform to the hypothesis of randomness. In 1962 Pathria tested and found no significant deviations among the first 10,000 digits of π. In 1965 R. G. Stoneham reported that the patterns noticed by Metropolis and his colleagues in the digits of e are not repeated beyond the first 2500 digits (up to 60,000). Stoneham tested 60,000 digits of e, claiming that the digits conformed to approximate randomness (except perhaps an excess of sixes!). In 1989, using a supercomputer built from spare parts in their New York City apartment, Gregory V. Chudnovsky and David V. Chudnovsky expanded π to over a billion digits, and their sophisticated random walk test suggests the billion digits are indeed random.[10]

There is, however, one issue that makes us uncomfortable in accepting the digits of π as random. In preparing to toss a coin or roll a die, or to measure the time intervals between clicks of a Geiger counter or an electronic pulse, we are satisfied that the random outcome will embody uncertainty. No one can know what is coming ahead of time. The idea of randomness is all bound up in our notion that the future is uncertain. The digits of π are another story. Although a particular individual may not

know, for example, the 81st digit of π, it is possible that *someone*
knows because the digits don't change—they are fixed. The digits are already there, perhaps waiting to be observed but not
waiting to happen.[11]

Criteria for Randomness

It was not until the twentieth century that mathematical definitions of randomness began to emerge. Beginning with a series of
lectures in 1919 and culminating in his 1928 landmark book on
a theory of probability, Richard von Mises attempted to provide
an intuitively satisfying definition of probability based on a
better understanding of randomness.

Von Mises defined randomness in a sequence of observations in terms of the inability to devise a system to predict *where*
in a sequence a particular observation will occur without prior
knowledge of the sequence. This definition is similar to the
impossibility of an unfair betting system: While a gambler may
follow a rule for when and how to bet, there exists no system of
prediction that could enable a gambler to bet in such a way as to
change his long-run relative frequency of success. The inability
to gamble successfully epitomizes an intuitively desirable trait of
a random sequence—namely, its unpredictability. This inability
to predict was called by von Mises "the impossibility of a gambling system." Randomness guarantees that there is no gambling

system, no formula, no rule capable of identifying particular elements in the sequence.[12]

As you might expect, some rather heated discussion ensued over von Mises's new theory of probability. Some philosophers rejected his long-run frequency theory outright and attempted to make the "commonsense" view of subjective probability more precise. They expounded a notion of probability based on our degree of knowledge or indifference, rather than long-run frequencies. Others, whom von Mises refers to as "nihilists," argued that definitions of probability and randomness were unnecessary and insisted that these concepts ought to be accepted as understood but undefinable mathematical terms. Interestingly, the largest group of von Mises's critics was the frequentists themselves.[13]

Von Mises addressed these criticisms in the 1939 edition of *Probability, Statistics and Truth*. The single largest criticism of his new theory was an objection to his definition of randomness. To say that prediction of the elements in a random sequence was impossible through *every* possible system, formula, or rule means that there could be *no* rule which could predict any substantial part of the sequence. Yet certainly every sequence conforms to *some* rule—we simply may not know what the rule is ahead of time. If every sequence conforms to some rule, a sequence which conforms to *no* rule could not exist. And if, quite by accident, we happen upon that rule or system, we would be able to change our probability of successful prediction. Many attempts were made to improve von Mises's definition of a random sequence by weakening the condition of *all possible* rules.[14]

In a 1963 paper Andrei Kolmogorov was able to show that if only *simple* formulas, rules, or laws of prediction are allowed, then von Mises-type sequences would exist. Kolmogorov hypothesized that "there cannot be a *very large number of simple algorithms.*" In 1965 Kolmogorov defined what he had meant by a "simple" law or formula and published a new quantitative measure of "information." In other words, rather than requiring the randomness of a sequence to be judged by absolute unpredictability, Kolmogorov would require only unpredictability by a small set of simple rules.[15]

Working in the field of information theory, Kolmogorov made further inroads in the measure of disorder by devising a method to quantify the complexity of an amount of information by the shortest length formula that could generate it. Modern mathematical definitions of random sequences have been constructed based on Kolmogorov's definition of complexity in information theory: a random sequence is one with maximal complexity. In other words, a sequence is random if the shortest formula which computes it is extremely long.[16]

His definition is intuitively appealing, for it conveys the idea that a random sequence cannot be described in any concise way—that is, no simple law can describe the sequence. Many sequences that are periodic *can* be described concisely, and Kolmogorov proposed that the absence of periodicity is, to common sense, a characteristic of randomness. The ultimate random sequence in terms of this complexity measure would be one that could be described only by naming the sequence itself, element by element; no formula that is shorter than the length of the sequence could be developed.[17]

Because this definition based on informational complexity measures the degree of disorder, perhaps different sequences can possess different degrees of randomness. In fact, Gregory Chaitin has pointed out that this new definition establishes a hierarchy of degrees of randomness. If randomness is thought of as an absolute, the longer the minimal-length formula which generates it, the more random the sequence. The problem is, How do we ever know if we have found the *shortest* formula?[18]

These three positions about the meaning of randomness—that it is a function of long-run frequency, a function of our ignorance, or a function of the length of its generating formula—are not irreconcilable. Upon close examination, we see a common, albeit unsurprising, element. Common to all of these views is the *unpredictability* of future events based on past events.

The frequentist von Mises described the randomness in a sequence as creating the impossibility of an unfair gambling scheme—no one knows what is coming. Randomness defined as an expression of man's ignorance or indifference certainly encompasses our inability to predict regardless of past knowledge. Randomness defined in terms of complexity, as the length of the shortest program which can describe it, ensures that the elements in a random sequence cannot be predicted in any concise way. Randomness guarantees us that no scheme could ever be conceived which would permit the probability of "knowing" to be improved.

In questions of physical randomness, such as the final distribution of billiard balls, whether there is some intrinsic uncertainty or only ignorance of complicated, confounding conditions, the random outcome is unknown and unknowable. Gustav Fechner believed that it was the novelty of each event—the fact that no moment or event is an exact replication of another—that introduces an amount of indeterminism into each new situation.[19]

Controversy still exists as to whether it is the outcome or the process which should determine randomness, that is, whether randomness is a characteristic of the arrangement itself or the process by which the arrangement was created, or both. In *The Logic of Chance,* Venn said that it is "the nature of a certain ultimate arrangement," and not "the particular way in which it is brought about," that should be considered when judging a random arrangement, and the arrangement must be judged by what would be observed in the long run. Basically, he seems to characterize randomness in terms of the disorderliness of the arrangement itself, not the process that generated it. But Venn then goes on to say that if the arrangement is too small, we must evaluate the nature of the agent that produced it, and often times the agent must be judged from the events themselves.[20]

Others, past and present, contend that "randomness is a property, not of an individual sample, but of the process of sampling," and thus nonrandom-looking sequences can be generated by a random process. Ian Hacking argues that "random samples are defined entirely in terms of the sampling device." Peter Kirschenmann believes a clear distinction should be made as to whether one is talking about random arrangement or random generation—in other words, the output or the process. Similarly, G. B. Keene distinguishes between a random sequence and a sequence chosen at random. Nicholas Rescher stresses that "randomness properly speaking characterizes sequences and the selection processes by which sequences can come about."[21]

Several authors have pointed out that a sequence obtained by a *perfect* random selection process may be very unrandom in appearance; as Cicero pointed out long ago, even the unlikely event has a chance of occurring, and *will* occur in the long run. This could prove most impractical for some applications of random selection. For example, Hacking points out that an enormously large table of random digits might very well contain an extensive string of zeros. If it did not contain such a string, the table's randomness might be suspect. Yet if the table contained such a string, this portion of the table would be unsuitable for small samples of digits. Hacking says he would be dissatisfied to exclude the string, since one could be accused of arbitrarily tampering with the table's randomness. On the other hand, he would be loath to include the string, since its use would be unsuitable for many purposes.[22]

G. Spencer Brown illustrates this paradox with sequences of coin tosses. He would like to reject certain sequences—for example, all heads or all tails, alternating heads and tails, and so on—as not being disorderly enough. Spencer Brown is by no means the only one who would like to remove portions of randomly generated sequences which appear nonrandom. The more reasonable justification for such removal lies in the usefulness of the subsequences in practical applications, however, not in how orderly they might appear. He further claims it is neither the sequence nor the process which defines randomness but rather the observer's psychology of disorderliness: we cannot know what is coming based on what we have already seen. Spencer Brown asserts, "Our concept of randomness is merely an attempt to characterize and distinguish the sort of series which bamboozles the most people . . . It is thus irrelevant whether a series has been made up by a penny, a calculating machine, a Geiger counter or a practical joker. What matters is its effect on those who see it, not how it was produced."[23]

Random sequences are commonly perceived as those which display no particular order. Venn observed that rather than the observer's state of mind causing the disorderly appearance of a random arrangement, it is the utter disorder of the arrangement itself which causes our uncertainty.[24] Part of the problem in addressing the question of whether or not absolute randomness exists stems from the myriad attempts to try to reconcile the scientific notion with the common notion of disorderliness.

Children have difficulty accepting that, in a chance experiment, a regular-looking sequence is as likely as an irregular-looking one. Age does not seem to solve this problem. In fact, several studies have shown that naive assessments of situations as random or nonrandom are often erroneous; even adults have mistaken notions about what randomness looks like. Mathematically unsophisticated subjects intuitively believe that random

outcomes ought to show variability from trial to trial—probably
indicating a desire for disorderliness. Even among sophisticated
subjects, researchers have shown perceptions of randomness to
be erroneous in expecting unjustified irregularity.[25]

The practical definitions of randomness—a sequence is random by virtue of how many and which statistical tests it satisfies
and a sequence is random by virtue of the length of the algorithm necessary to describe it—are not without their deficiencies. Statistical tests of randomness will always involve error.
Error causes the statistical test to reject as nonrandom some
sequences that are indeed random and to accept as random
some sequences that are nonrandom. The significance level of
the test used will control the probability and the percent of
error, but the error can never be reduced to zero; it is integral to
the calculus of probability.

For example, we may find that there is a 1 in 1000 chance
that our sequence does not exhibit a specific (nonrandom) behavior when tested. This does not mean that the nonrandom
behavior is not there; it means that it is highly unlikely to be
there. In the long run, we expect to be wrong 1 time in 1000.
The significance level of the test is 1 in 1000, and the chance
that we are in error in judging our sequence to have passed the
test is 1 in 1000. Furthermore, this error can never be reduced
to 0 because, in the long run, even the unlikely will happen. If a

sequence had passed all available tests, it might only mean that the tests used were not sensitive to the particular regularity that existed. Karl Popper has said that we have no tests for the presence of all regularity, only ones for "some given or specific regularity."[26]

A similar hitch resides in the complexity threshold level against which a sequence will be deemed random—contingent on the length of its producing agent (algorithm). The length of the shortest program which can produce a maximally complex sequence is equal to the length of the sequence itself. Instead of requiring maximal complexity, if the sequence is complex enough, we may have to judge the sequence random enough.

Opposing positions are held as to whether absolute randomness is possible and whether there can be degrees of randomness. Some view absolute randomness as a limit concept; absolute randomness is the ideal, though perhaps unobtainable.[27] Others declare that there is no absolute randomness, only relative randomness.

Kendall and Babington-Smith, for example, have argued for *local* randomness and have proposed four tests to assess it: the frequency test, serial test, poker test, and gap test. The *frequency test* is a test of the uniform occurrence of each of the ten digits 0 through 9—our expectation being that each digit will occur an approximately equal number of times, about one time in ten.

The *serial test* is a test of the uniform occurrence of two-digit pairs—we expect that each possible two-digit pair occurs an approximately equal number of times, about one time in one hundred. Why should we test to see if each digit occurs about one-tenth of the time *and* to see if each two-digit pair occurs about one-hundredth of the time? Consider the sequence 0 1 2 3 4 5 6 7 8 9 0 1 2 3 . . . Although it does not appear to be a random sequence, each digit will occur about one-tenth of the time (depending on where we stop in the sequence). However, if *two*-digit pairs are considered—01 23 45 67 89 01 23 . . .—the pattern is revealed, since only 5 pairs out of 100 pairs will ever appear.[28]

The *poker test* compares five-digit blocks against the expected occurrence of certain five-card poker hands: five-of-a-kind, four-of-a-kind and one other, three-of-a-kind and two others, two pairs, one pair, and all different digits. The *gap test* examines the number of digits (or gap length) between the occurrences of the digit 0. For example, the sequence 0|12|0|0|924789|0 has gaps of length 2, 0, and 6, respectively. The frequencies of the lengths of the gaps between successive zeros in the sequence is compared against what would be expected of digits selected by chance.[29]

Their tests were clearly aimed at determining the usefulness of the particular sequence of random digits for sampling. And while a sequence might be determined to be useful for one purpose, it might not be useful for another. In particular, the longer the sequence the more likely it is to have bad patches which would not be locally random in themselves.

One author has pointed out that we can have relative randomness in the same way that we can have relatively accurate measurement without having knowledge of absolutely accurate measurement.[30] For example, with progressively more sophisticated measuring devices, we can measure the width (in inches) of a sheet of paper as 8.5, 8.501, 8.50163, and so on. At what point do we stop to pronounce one result the correct measurement? Perhaps we can never have the exact measurement, but at some point we proclaim the accuracy as being close enough.

Is nearly random good enough? Kolmogorov thought that only approximate randomness applied to finite populations.[31] A perfectly random sequence may be merely an abstraction, but approximate randomness may be good enough. Of course, we must add here that the definition of "good enough" may depend on the situation. A casual mixing of tickets to draw a door prize may be random enough, while we may expect more care to be taken in the mixing of the Selective Service draft tickets for induction during a war.

As the use of random sampling in statistical experimentation became the norm, more tables were generated and more tests

developed. Yet since the first table of random digits was pub-   **173**
lished by L. H. C. Tippett in 1927, there has been controversy
over how to produce random sequences of digits and how to test
the sequences for true randomness. Statistical tests to validate
(or invalidate) the randomness of a sequence have been devised,
and there is no limit to the number of tests that could be
conceived.[32]

Though some tests have become standard, it remains ques-
tionable which tests and how many of these tests a sequence
must satisfy to be accepted as random. John von Neumann
stated that in his experience it was more trouble to test random
sequences than to manufacture them.[33]

Paradoxes in Probability

10

Paradoxes occur everywhere in mathematics, but most often they occur at a rather sophisticated level. In probability, however, paradoxes and counter-intuitive situations occur at a compara-

tively simple level. Perhaps this is one reason why people's intuition for probability is not as keen as their intuition for geometry or arithmetic. The renowned mathematics educator George Polya has pointed out that intuition comes to us naturally, and that formal (mathematical) arguments should legitimize this intuition. But educators who have studied the difficulty of teaching probability point out how counter-intuitive even simple probabilities can be.[1]

To take an example, most of us would consider it quite extraordinary to meet someone sharing our birthday. Yet among a group of 25 or more persons, the chances are better than 50-50 that two or more people will have the same birthday. The reason that this result is so surprising is that we all tend to focus on a *particular* birthday (usually our own). We think the question is: What is the probability that one or more of the other people in this group has *my* birthday? Indeed the probability of that occurring (if there are 25 people in the group) is about 0.064, or less than 7 percent—not anywhere close to 50 percent. But when the question is not about a particular person or a particular birthday but about any two or more people sharing any birthday, the chances are indeed better than 50 percent.

Another classic probability paradox is usually stated in the following way:[2]

> Given that a family has two children and at least one is a girl, what is the probability that the family has two girls?

The usual assumption is that in the birth of any one child, the birth of a girl and the birth of a boy are equally likely. This particular problem is paradoxical, however, because stories can be constructed that seem to change your knowledge very little, yet change the answer to the question completely. Consider the problem when worded this way:

> (1) You make a new friend, and you ask if she has any children. Yes, she says, two. Any girls? you say. Yes, she

says. What is the probability that both are girls? Answer: one-third.

Like the toss of two coins that can result in the equally likely outcomes of HH, HT, TH, or TT, the birth of two children can result in the outcomes GG, GB, BG, or BB. Since we know that there is at least one girl, the last situation (BB) is impossible. So, GG (both girls) is one outcome out of the three possible equally likely outcomes; the probability is ⅓.

Now consider the same problem with the story constructed this way:

(2) You make a new friend, and you ask if she has any children. Yes, she says, two—ages 6 and 10. Is the oldest a girl? you ask. Yes. What is the probability that both of her children are girls? Answer: one-half.

The question now is really: If your first child is a girl, what is the probability that your second will be a girl? That probability is the same as the probability of the birth of a girl, namely ½. Looked at another way, the possible outcomes for the birth of two children, *in birth order,* are now known to be only GG or GB, both of which were equally likely before we obtained the additional information about the gender of the eldest. They are still equally likely. Since GG is one instance of the two possible

equally likely instances, the probability of two girls is 50 percent.

And finally, consider this rendition of the problem:

> (3) You make a new friend, and you ask if she has any
> children. Yes, she says, two. Any girls? Yes. The next day
> you see her with a young girl. Is this your daughter? you
> ask. Yes, she says. What is the probability that both of her
> children are girls? Answer: one-half.

This seems so peculiar, because it doesn't appear that we have
any more information than we did in the first example, yet the
probability is different. Before we saw her for the second time,
we already knew that one of her children was a girl, and at the
second meeting we learn nothing new about birth order. But
once again, the question has changed. The question is now:
What is the probability that the child you don't see is a girl? And
that probability is the same as the probability of the birth of a
girl, $\frac{1}{2}$. In other words, the possible outcomes are that you see a
girl and don't see the other girl (GG), or you see a girl and you
don't see the other child, a boy (GB). Since two girls is again one
outcome out of two possible equally likely outcomes, the probability of two girls is $\frac{1}{2}$. The answer you get depends on the
story told; it depends on how you came to know that at least
one child was female. Considering how much the language of

the problem affects the answer in this simple example, it is no wonder that probability is a perplexing science to many people.

Let's next look at the Jailer's Paradox:[3]

> Adam, Bill, and Charles are being held in a prison. The jailer is the only one who knows which of the three is condemned to death and which two will go free. Adam, who has a probability of ⅓ of being executed, has written a letter to his mother and wants it to be delivered by either Bill or Charles, whichever one gets to go free. When Adam asks the jailer whether he should give his letter to Bill or to Charles, the jailer, a compassionate man, feels he is faced with a dilemma. If he tells Adam the name of the man who will go free, he thinks, then Adam now has a probability of ½ of being condemned to death, since either Adam or the remaining man must be put to death. If he withholds the information, Adam's chances remain at ⅓. Since Adam already knows that one of the other two men will go free, how can his chances of being executed be affected by his knowing the man's name?

The short answer is that Adam's chances don't change, and the jailer is mistaken to worry. Even if Adam is given the name of the man who will go free, Adam will still have a ⅓ chance of

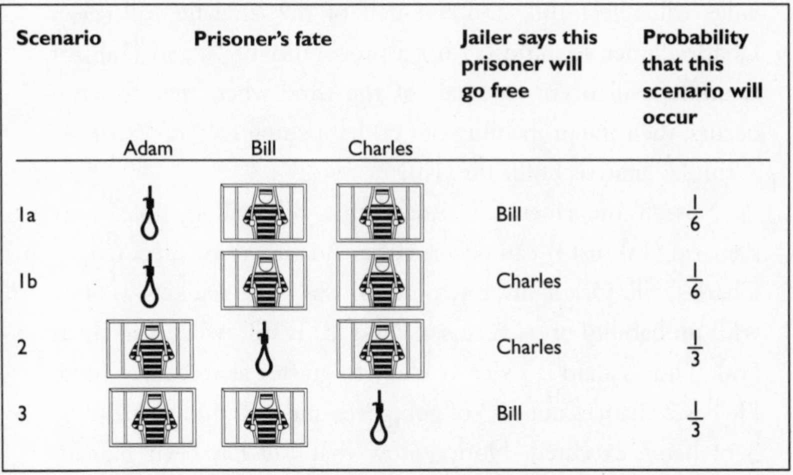

Scenario	Prisoner's fate			Jailer says this prisoner will go free	Probability that this scenario will occur
	Adam	Bill	Charles		
la				Bill	$\frac{1}{6}$
lb				Charles	$\frac{1}{6}$
2				Charles	$\frac{1}{3}$
3				Bill	$\frac{1}{3}$

FIGURE 27 The jailer's paradox: If he divulges the name of one of the two prisoners who will go free, will that change the probability of being executed for the remaining two?

being executed. The man who was not named, however, now has a $\frac{2}{3}$ chance of being executed! How can that be?

Initially, each of the scenarios (1), (2), and (3) are equally likely, each with a probability of $\frac{1}{3}$ (see Figure 27). If (1) is the actual scenario (that is, Adam has been condemned to death), then (1a) and (1b) are equally likely, since the jailer has the choice to name either Bill or Charles. One-half of the time the jailer will select Bill, and one-half of the time he will select Charles. Since scenario (1) has a probability of $\frac{1}{3}$ and (1a), for example, will occur one-half of the time when that scenario occurs, then the probability of (1a) happening is $\frac{1}{2}$ of $\frac{1}{3}$, or $\frac{1}{6}$. A similar analysis holds for (1b).

Now if the jailer tells Adam that Bill will go free, only scenario (1a) or (3) can occur: either Adam will be executed, or Charles will. Originally, outcome (3) was twice as likely as (1a), with probability of $\frac{1}{3}$ versus $\frac{1}{6}$, and (3) is still twice as likely as (1a). Thus, Adam is twice as likely to go free as to be executed. He has 2 chances out of 3 of going free and only 1 chance out of 3 of being executed. Notice, now that Bill has been named, Charles is twice as likely to be executed.

Under scenarios (2) and (3), the jailer cannot choose who to name. This phenomenon is known as restricted choice and is familiar to bridge players. Restricted choice played a part in the now-infamous Monty Hall Problem which appeared in the September 1990 *Parade* column, "Ask Marilyn." The problem is named after the host of *Let's Make a Deal,* Monty Hall. Marilyn vos Savant is asked the following question by a reader:

Suppose you're on a game show, and you're given the choice of three doors: Behind one door is a car; behind the others, goats. You pick a door, say No. 1, and the host, who knows what's behind the other doors, opens another door, say No. 3, which has a goat. He then says to you, "Do you want to pick door No. 2?" Is it to your advantage to take the switch?

Her answer: You should switch. According to the *New York Times,* the problem and her solution were "debated in the halls of the Central Intelligence Agency and the barracks of fighter pilots in the Persian Gulf" and "analyzed by mathematicians at the Massachusetts Institute of Technology and computer programmers at Los Alamos National Laboratory in New Mexico."[4]

Like the jailer, Monty (in the problem as posed) has restricted choice under certain scenarios (see Figure 28). Originally, your chances of picking a door with a car behind it were one out of three, or $\frac{1}{3}$. Just as Adam's chances of being executed did not change by being informed of the name of one of the other two prisoners to go free, your chance of having selected the door with the car behind it has not changed now that you have seen a goat. Therefore, if you stick with your choice, your chance of getting the car is $\frac{1}{3}$, whereas if you switch after one door has been eliminated, your chances of getting the car are $\frac{2}{3}$.

The next paradox is the case of the Intransitive Spinners.[5] Consider a game between two players, each with a spinner that, when spun freely, can point to one of two equally likely num-

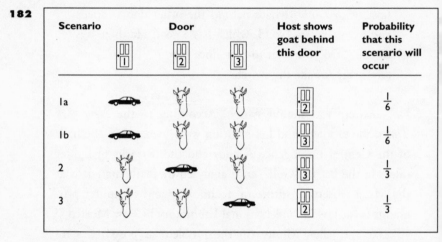

FIGURE 28 The Monty Hall problem: After being shown a goat behind one of the two unpicked doors, should the contestant stick with his initial choice or switch?

bers; the player who spins the highest number wins. There are three players available: Annie, who has a spinner numbered 8 and 4; Betsy, who has a spinner numbered 10 and 0; and Carla, who has a spinner numbered 6 and 2 (see Figure 29).

If Annie plays Betsy, each has a probability of $\frac{1}{2}$ of winning, since Betsy always loses with a 0 and always wins with a 10. Annie's spinner can spin a 4 or an 8, but it doesn't really matter. What matters is what Betsy spins. If Betsy plays Carla, whose

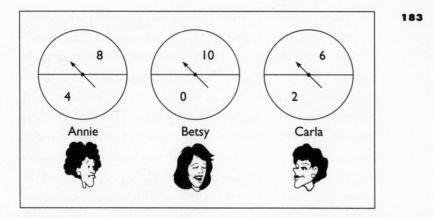

FIGURE 29 The problem of the Intransitive Spinners: If Annie and Betsy are equally matched, and Betsy and Carla are equally matched, does it follow that Annie and Carla are equally matched?

spinner can spin a 2 or a 6, each has a $\frac{1}{2}$ probability of winning; again, Betsy always loses with a 0 and always wins with a 10. Annie and Betsy are equally matched, and Betsy and Carla are equally matched. So can we conclude that Annie and Carla are equally matched?

Intuitively, we may feel that the correct answer is yes. But it is no. If Annie plays Carla, there are four possible equally likely outcomes. If the spins are 8 and 6, Annie wins; 8 and 2, Annie wins; 4 and 2, Annie wins; 4 and 6, Carla wins. Carla can win only if she spins 6 and Annie spins 4. The probability that Annie will beat Carla is $\frac{3}{4}$. Annie and Carla are not equally matched.

The final example of the counter-intuitive nature of probability is called Simpson's Paradox.[6] It was discovered that a college was accepting women at a lower rate than they were accepting men. The administrators wanted to find out if there was one department responsible for distorting the total statistics. In the course of their investigation, they gathered acceptance data from each department in the school. Expecting to find the culprit department that was tarnishing their image, they discovered instead that in every single department, the acceptance rate for females was *higher* than that of males. It seemed as though some information must have been missing or miscounted. If every single department is counted only once and there is no overlap, how is it possible to have a higher female rate in every single department and a lower rate overall?

Let's suppose that the college has an acceptance rate of $\frac{50}{90}$, or about 56 percent, for females compared to $\frac{60}{100}$, or 60 percent, for males and has two departments (see Figure 30). In Department 1, 50 women apply and 20 are accepted; 30 men apply and 10 are accepted. The $\frac{20}{50}$, or 40 percent, acceptance rate for females compares quite favorably to the $\frac{10}{30}$, or about 33 percent, acceptance rate for males. In Department 2, 40 women apply and 30 are accepted; 70 men apply and 50 are

Department	Applied ♀	Accepted ♀	Applied ♂	Accepted ♂
1	50	20 (40%)	30	10 (33%)
2	40	30 (75%)	70	50 (71%)
College Total	90	50 (56%)	100	60 (60%)

FIGURE 30 Simpson's Paradox: If the acceptance rate of women is higher than the rate of men in every department in the college, how can the women's rate be lower than the men's for the college as a whole?

accepted. The acceptance rate for females is $^{30}/_{40}$, or 75 percent, compared with the acceptance rate for males of $^{50}/_{70}$, or about 71 percent. Yet when the two sets of statistics are combined, the college acceptance rate for women, $^{50}/_{90}$, is smaller than the $^{60}/_{100}$ rate for men.

It seems to me that the greatest impediment to the development of probability has been the absence of an understanding of equally likely outcomes in all but the simplest events, combined with a superstitious belief in fortune or luck. There is evidence that for a single lot or die, however many sided, the notion of equiprobable was well understood from early times. With the astragalus, however, which does not have equally likely sides, and in games employing two or more dice, the concept of equally likely outcomes is highly obscured. Although some ancient dice were quite well made and true, without a great deal of experience or a keen intuition, it is extremely difficult to identify the equally likely outcomes of a compound event like the throw of two or three six-sided dice.

This misunderstanding, coupled with a misconception about probabilities in the short run versus the long run, encouraged a belief in streaks of good and bad luck, possibly brought on by a deity. The ancient Greeks seem to have concluded from their encounters with chance that precision and law resided only in the divine realm, and chaos and irregularity characterized the world of man. Being unable to reconcile their idealized notions of natural law with the evidence of an imperfect physical world, they failed to develop a science of probability.[7]

But during the Middle Ages, though ideas of chance were still far from precise, correct notions about probability were not impossible for the medieval mind to grasp. The historian Edmund Byrne points out that people in the Middle Ages, like us, "were in daily contact with the puzzling uncertainties of the contingent, the accidental, the chance event."[8] Why did theories about games of chance, frequency, randomness, and probability appear much later?

Numerous explanations for this slow evolution have been

offered by some and rebutted by others. They include the impi-
ety of inserting chance into a divine decision; obsession with
determinism and necessity; the lack of empirical examples of
equiprobable alternatives; the absence of an economic problem
that the science of probability satisfied; the absence of an ade-
quate numerical notation system; the absence of combinatorial
algebra; the superstition of gamblers; and moral or religious
barriers, particularly in the Christian Church, where it was believed that everything, however small, happened at the direction
of God—"that even the hairs of the human head are numbered,
and not a sparrow falls without God's knowledge."[9]

M. G. Kendall points out that "it seems to have taken
humanity several hundred years to accustom itself to a world
wherein some events were without cause; or, at least, wherein
large fields of events were determined by a causality so remote
that they could be accurately represented by a non-causal model.
And, indeed, humanity as a whole has not accustomed itself to
the idea yet. Man in his childhood is still afraid of the dark, and
few prospects are darker than the future of a universe subject
only to mechanistic law and to blind chance."[10]

In the short run, chance may seem volatile and unfair. And
while experience with long-run frequencies can help to mod-
ify some of our maladaptive behaviors based on a misunderstanding of randomness and probability, a *very* long run may

be required. Considering the misconceptions, inconsistencies, paradoxes, and counter-intuitive aspects of probability, it should be no surprise that, as a civilization, we took a long time to develop correct intuitions. Indeed, every day we can see evidence that the human species does not yet have a very highly developed probabilistic sense. Perhaps we all should approach chance encounters with caution, in the short run.

Notes

Bibliography

Index

Notes

1 Chance Encounters

1. Burrill, 1990 (p. 117).

2. Tversky and Kahneman, 1982 (p. 156).

3. Nisbett, Borgida, Crandell, and Reed, 1982 (p. 111).

4. Cassells, Schoenberger, and Grayboys, 1978 (p. 999). In this case we are assuming that the test correctly diagnoses the disease every time; that is, there are no false negatives.

5. The numbers reported for the accuracy of the Mantoux skin test vary with the prevalence of TB in the population being tested. These are estimates for illustrative purposes and were based on Remington and Hollingworth (1995) and the American Academy of Pediatrics Committee on Infectious Diseases (1994). Estimates of incidence of TB infection in the population were based on conversations with epidemiologists at the Center for Disease Control and Prevention.

6. Kahneman and Tversky, 1982 (p. 32).

7. Piaget and Inhelder, 1975 (p. 212).

8. Tversky and Kahneman, 1974 (p. 1130).

9. Abraham De Moivre, 1756.

2 Why Resort to Chance?

1. Goldstein, 1971 (p. 172). Goldstein paraphrased the various responses, and noted that the percentages add to more than 100 percent due to multiple responses.

2. Jackson, 1948.

3. Sanhedrin 43b, in Epstein, 1935. See Hasofer, 1967.

4. The story of the excavations at Ur can be found in Woolley, n.d., and Woolley, 1928. A description of the archaeological finds can also be found in "Ancient die," 1931. Evans, 1964, describes the finds at Knossos.

5. See Carnarvon and Carter, 1912; Petrie and Brunton, 1924; and Gadd, 1934, for descriptions of the boards found at Babylonian, Assyrian, Palestinian, and other sites. The board game from an Egyptian tomb in Thebes is on permanent display in the Metropolitan Museum of Art in New York City; its excavation is described in Carnarvon and Carter, 1912.

6. The die found in excavations in Northern Iraq is described in "Ancient die," 1931. See also Mackay, 1976, and Bhatta, 1985. That pips were used because there was no numerical notation system was suggested by Davidson, 1949, and David, 1962.

7. See Smith, Wayte, and Marindin, 1901; Pease, 1920; David, 1962; Hasofer, 1967; *OED*, 1989; Culin, 1896.

8. "News and views," 1929; Falkener, 1892.

9. See Quibell, 1913; Evans, 1964; Petrie and Quibell, 1896. Culin (1896 and 1907) has reported two-sided "dice" made from beads or shells that were used as chance mechanisms in divination practices and games by Native Americans and Africans.

10. Smith, Wayte, and Marindin, 1901; Pease, 1920; Suetonius, *Aug.* 13.

11. Suetonius, *Aug.* 71. See Smith, Wayte, and Marindin, 1901.

12. Winternitz, 1981 (p. 103), quotes a translation by A. A. Macdonell, *Hymns from the Rgveda* (pp. 88ff).

13. See B. Walker, 1968; Hacking, 1975; van Buitenen, 1975.

14. See Tylor, 1879; B. Walker, 1968. De Vreese, 1948, analyzes the rules of the vibhitaka game.

15. *Mahabharata* 3(32)70.23, trans. van Buitenen, 1975. Hacking, 1975 (p. 7), relates a different translation, by H. H. Milman, "I of dice possess the science and in numbers thus am skilled."

16. De Vreese, 1948.

17. Grierson, 1904; Hacking, 1975 (p. 7).

18. Suggested by de Vreese, 1948 (p. 362). See also Tylor, 1873; Held, 1935.

19. Culin, 1907 (p. 45).

20. Tylor, 1896; Culin, 1896, 1903, 1907; Erasmus, 1971.

3 When the Gods Played Dice

1. Homer, *Iliad,* 3.314ff.

2. Ibid., 7.171ff.

3. Piaget and Inhelder, 1975 (p. 7).

4. Tacitus, *Germania* 10.

5. Cicero, *De div.* 2.51.85–87. See Pease, 1920.

6. Maimonides' quotation is from Hasofer, 1967 (p. 40); see also Rabinovitch, 1973.

7. Biblical references are from the King James Version of *The Holy Bible:* Levit. 16:8; Esth. 3:7, 9:24ff; I Sam. 10:20; I Chron. 24:5ff, 25:8ff, 26:13ff; Judges 20:9; Neh. 10:34, 11:1; I Sam. 14:41ff; Jonah 1:7; Jos. 7:16ff; I Sam 10:20ff; Jos. 7:16ff.

8. See Num. 26:53ff, 33:54, 34:13, 36:2; Jos. 14:2, 15:1, 16:1, 17:1ff, 18:6ff, 19:1,10,17, 21:4ff; I Chron. 6:39ff; Joel 3:3; Nahum 3:10; Neh. 3:10; Ps. 22:18.

9. Lichtenstein and Rabinowitz, 1972; Oppenheim, Gelb, and Landsberger, 1960.

10. Hasofer, 1967.

11. For a full description, see Fienberg, 1971. There were 365, not 366, birthdates in 1951, the birth year of the men in the second lottery.

12. Lichtenstein and Rabinowitz, 1972.

13. Hasofer, 1967.

14. Oppenheim, 1970.

15. See, for example, Ashton, 1893; Ewin, 1972; Sullivan, 1972; Daston, 1988.

16. The writing of the basic text of the *I Ching,* which was completed no later than the seventh century B.C. (and some attribute its beginnings to 3000 B.C., others to 1140 B.C.), was probably preceded by a system of divination from which the text took shape in the oral tradition from agricultural folklore, according to Shchutskii, 1979, and Cleary, 1986.

17. Fisher-Schreiber et al., 1989.

18. Wilhelm, 1950, thinks that a *Book of Changes,* based on trigrams, existed during the Hsia dynasty (2205–1766 B.C.), and another is mentioned by Confucius as existing during the Shang dynasty (1766–1150 B.C.). In fact, Tibetan Buddhists use trigrams, not hexagrams, for divina-

tion, according to Hastings, 1912, and Waddell, 1939. See also Shchutskii, 1979, and Fisher-Schreiber et al., 1989.

19. For a thorough description of the methods of casting an oracle, Wilhelm's 1950 translation (or an English translation of Wilhelm's German) is one of the best, according to Shchutskii, 1979, and Fisher-Schreiber et al., 1989. Wilhelm's is a clear presentation of the two methods of casting the oracle.

20. Wilhelm, 1950.

21. Shchutskii, 1979 (pp. 57, 232–233).

22. Ibid.

23. Wilhelm, 1950.

24. Cicero, *De div.* 2.110.

25. For a discussion, see Parke, 1988. Virgil describes the Sibyl's oracles as "the mystic runes you utter" (*Æneid* 6.72). Cicero, in describing the oracle at Praeneste (which was not one of the Sibyls), mentions oracles written in "ancient characters" (*De div.* 2.85).

26. *Æneid* 3.441–452.

27. The Greek manifestation was Tyche and the Assyrian and Babylonian goddess was Ishtar (Oppenheim, 1970; Langdon, 1930, 1931). For a thorough treatment of Fortuna, see Patch, 1927.

28. Pease, 1920 (p. 373); Chaucer, 1949 (pp. 86, 93); William Blake, 1825–1827.

4 Figuring the Odds

1. David, 1962 (p. 22). The probabilities might also be affected by the surface the bones were tossed upon.

2. Galileo, 1623.

3. De Moivre, 1756 (p. 3).

4. Kendall, 1956; David, 1962.

5. Kendall, 1956.

6. According to Kendall, 1956, this work was chronicled by Baldericus in the eleventh century and not published until 1615.

7. Kendall, 1956.

8. Cardano, 1564; Ore, 1953; David, 1962.

9. Galileo, 1623.

5 Mind Games for Gamblers

1. Todhunter, 1865; Kapadia and Borovcnik, 1991.

2. The psychologists Tversky and Kahneman (1971) label this misconception "belief in the law of small numbers" and state that this belief causes "exaggerated faith in the stability of results observed in small samples."

3. Buffon's essay (1777) was a supplement to his *Histoire naturelle*. The paradox had come to the attention of mathematicians through Nicholas Bernoulli and his brother, Daniel, who held posts at the Petersburg Academy.

4. Boyer, 1968, posed the Petersburg problem as the game of "Peter and Paul."

5. Buffon, 1777; Todhunter, 1865.

6. Weaver, 1963; Tversky and Bar-Hillel, 1983.

7. Kolata, 1990.

8. Diaconis and Mosteller, 1989 (p. 859).

9. Cicero, *De div.* 2.21.480, 2.59.121.

10. Jonah 1:7. See Rabinovitch, 1973 (p. 28).

11. Cardano, 1564 (pp. 189, 204, 223).

12. Ibid. (pp. 192, 196). Shafer, 1978, points out that Cardano's understanding of equally likely outcomes is carried to the extreme when he calculates probabilities for the talus as if each of the nonsymmetric sides was equally likely. See also Ore, 1953; Hacking, 1975.

13. Piaget and Inhelder, 1975 (p. 77).

14. Tversky and Kahneman, 1974 (p. 1125).

15. Vos Savant, December 26, 1993 (p. 3), quoted in CHANCE website.

16. Ore, 1953; Galileo, 1623.

17. Ore, 1960; Boyer, 1968.

6 Chance or Necessity?

1. Leucippus fragment (67.B.2) from H. Diels, *Die fragmente der Vorsofratiker,* 6th ed. (Berlin, 1951), quoted in Sambursky, 1959a (p. 50). On Democritus see Cioffari, 1935, and Sambursky, 1959b.

2. Quoted by Plutarch, *De Stoic, repugn.* 1045 c; see Sambursky, 1959a (p. 56).

3. Hobbes, 1841; Sheynin, 1974.

4. Hobbes, 1841 (pp. 41–42).

5. Hobbes, 1841 (p. 413).

6. Arbuthnot, 1714.

7. Porter, 1986 (p. 233).

8. See Simpson, 1756; Sheynin, 1971. The problem of personal observer errors wherein one observer may be consistently higher or lower (faster or slower) than another is seen in the story of Nevil Maskelyne, the fifth Astronomer Royal, and his unfortunate assistant, David Kinnebrook. Maskelyne, who managed the Royal Observatory for 46 years, eventually fired Kinnebrook believing him to be lazy and incompetent. Ronan, 1967, in *Astronomers Royal,* says that Kinnebrook was wrongfully blamed when, through no fault of his own, he consistently observed a star transit later than Maskelyne himself.

9. Galileo, 1632. See Maistrov, 1974.

10. Kendall, 1961; Maistrov, 1974.

11. Simpson, 1756; H. Walker, 1929; Kendall, 1961.

12. D. Bernoulli, 1777 (pp. 158, 165). I find the example of the chance deviation in an archer's aim interesting. According to the *OED,* the modern word "stochastic" (meaning "randomly determined; that follows some random probability distribution or pattern, so that its behaviour may be analysed statistically but not predicted precisely") is derived from the Greek *"stochastikos,"* meaning "skillful in aiming, proceeding by guesswork," derived from *"stochos,"* meaning "aim, guess" (*OED,* 1989).

13. On Laplace see H. Walker, 1929; Stigler, 1978. See also Adrian, 1808; Gauss, 1809.

14. Laplace, 1886.

15. Stigler, 1986.

16. Weldon, 1906; Porter, 1986; H. Walker, 1929.

17. K. Pearson, 1924; De Moivre, 1756.

18. K. Pearson, 1924; H. Walker, 1929.

19. A quincunx is an arrangement of five objects at the four corners of a rectangle and one in the center. Based on the design on the face of an ancient Roman coin, it was an arrangement used in planting trees. See Stigler, 1989.

20. Ibid. It is interesting to note that Galton used the expression "deviated normally on either side of their own mean" (p. 513). This is the first instance I have come across of the use of the word "normal" in referring to the law of errors. Stigler, 1989, thinks that Galton never actually constructed the second *quincunx,* but that it was entirely conceptual.

7 Order in Apparent Chaos

1. Laplace, 1814 (p. 66). Quoted in Stigler, 1986. See also Kruskal and Mosteller, 1980; Kendall, 1941.

2. Fechner's work culminated in the manuscript, *Kollektivmasslehre,* which was a theory of frequency distributions. According to Michael Heidelberger, 1987 (p. 139), who brought that work to light, Fechner originated the concept of a "collective object," defining a collective as a sequence of specimens each varying randomly according to chance, independent of any law of nature.

3. Heidelberger, 1987 (pp. 136, 141).

4. DeForest, 1876; Stigler, 1978, 1991.

5. Galton, 1890b. See also Stigler, 1978, 1980.

6. Peirce and Jastrow, 1884.

7. E. Pearson, 1967 (p. 344). Edgeworth, 1885a, claims to have com-

200   puted 280 sums of 10 digits taken at random from mathematical or statistical tables, but he elaborates no further.

8. Edgeworth, 1885a (p. 205).

9. Edgeworth, 1885b (p. 637).

10. Weldon, 1906.

11. Darbishire, 1907 (p. 13).

12. K. Pearson, 1895. His son, Egon S. Pearson, 1965, stated that the device was possibly constructed for Karl Pearson's lectures on probability at Gresham College given in 1893. See K. Pearson, 1900 (p. 157).

13. K. Pearson, 1900 (p. 174).

14. Gosset had spent part of the year 1906–1907 at the Biometric Laboratory in London, as the "student" of Professor Karl Pearson. See E. Pearson, 1967.

15. Student, 1908a.

16. Gosset needed data that were approximately normally distributed and correlated. In a paper written in 1890, Galton had stated: "Thus we may speak of the length of the middle finger and that of the stature being correlated together under a recognized understanding that the variations are quasi-normal." See Galton, 1890a (p. 84).

17. Student, 1908b.

18. Bispham, 1923 (p. 693). Bispham's two experiments were designed to investigate the dispersion of the partial correlation coefficients in samples (of size 30) from an uncorrelated population and in samples (of size 10, 30, and 60) from a highly correlated population. In the first study, the draws were independent (and theoretically uncorrelated), and in the second the sums were not independent (and therefore theoretically correlated).

8 Wanted: Random Numbers

1. Tippett, 1925, describes these digits as "taken at random" from the middle digits of the parish areas in census reports. In 1927, when the digits were published in a table, Karl Pearson also described the digits as "taken at random" (Tippett, 1927). Neither scientist described precisely what procedures were used to ensure randomness.

2. Bork, 1967; Tippett, 1927 (pp. iii–iv).

3. Fisher and Yates describe their elaborate procedure only in the first edition of *Statistical Tables for Biological, Agricultural and Medical Research*, 1938. According to Kendall and Babington-Smith, 1939a, Fisher and Yates apparently used several tests on their data and, upon the discovery of an excess of sixes, they removed some of them and replaced them with other digits. Kendall and Babington-Smith themselves had originally tried to use the London telephone directory as a source of random numbers. However, having selected 10,000 of them, with great care to prevent a bias of any sort, they determined that there was a deficiency of numbers ending in 5 or 9. Therefore, the telephone directory would be a poor source of random digits. See Kendall and Babington-Smith, 1938, 1939a, 1939b; Peatman and Shafer, 1942; Vickery, 1939.

4. Horton, 1948. His method actually involved products of random digits, but a subsequent paper (Horton and Smith, 1949) proved that this was equivalent to addition modulo 2 and could be generalized to any base. Modulo 2 means that each number is replaced by its remainder after division by 2. For example, since 3 has a remainder of 1 when divided by 2, 3 is replaced by 1 (modulo 2); 4 is replaced by 0 (modulo 2); 5 is replaced by 1 (modulo 2), and so on. The ICC digits were produced from

202   a sequence of random digits by addition modulo 10 and punched onto computer cards.

5. RAND, 1955.

6. Buffon, 1777; Laplace, 1886. See also Maistrov, 1974; Boyer, 1968.

7. Meyer, 1956.

8. See Hall, 1873; Gridgeman, 1960; De Morgan, 1912; Clark, 1933, 1937. Others have indicated that there might have been outright cheating.

9. According to Rhodes, 1995, Ulam says that he came up with this idea while playing solitaire during his recovery from a serious illness. Ulam found that he could estimate the outcome of the game from examining his success in just a few trial cards. In the original von Neumann-Ulam sense, the Monte Carlo method was used in models of physical situations to solve a deterministic problem (like solutions to numerical integration equations) by finding a probabilistic analogue and obtaining approximate answers to the analogue problem by some experimental random sampling procedure. See Meyer, 1956; Report on Second NBSINA Symposium, 1948.

10. For this reason, the generation of digits by a physical random process, which could not be replicated, was considered unsatisfactory, according to von Neumann, 1951.

11. Von Neumann, 1951 (p. 768); Knuth, 1981.

12. Von Neumann, 1951.

13. Gardner, 1975 (p. 169); Knuth, 1981.

14. Generators that use the bit structure of computer-stored information are called shift-register generators; these have been criticized because of the burden of manipulating long strings of digits within the algorithm. See Marsaglia and Zaman, 1991.

15. The number 2,147,483,647 is the Mersenne prime, $2^{31} - 1$. Park and Miller, 1988, have stated that it is known that congruential generators suffer from the deficiency that, if the digits (Z_n, Z_{n+1}, Z_{n+2}) are graphed as points in three dimensions, the points fall on a finite, and possibly small, number of parallel planes rather than falling uniformly in an apparently random pattern. This is not a problem in most applications, but, if necessary, a very powerful test—the spectral test, formulated by R. R. Coveyou and R. D. MacPherson in 1967, can detect deficiencies in higher dimensions; see Knuth, 1981. Marsaglia and Zaman, 1991, have also objected to the high cost of arithmetic modulo a prime number in congruential generators.

16. Marsaglia and Zaman, 1991, have been able to show that their generators have tremendously long periods, that is, the digit cycle is long. See Peterson, 1991b.

17. The obvious problem with this generator is that an even number always follows two odds (the sum of two odd numbers is even), and an odd number always follows an even and an odd (the sum of an even number and odd number is odd).

18. Knuth, 1981 (p. 5); this massive survey of generators and tests has become a classic work. See also Coveyou and MacPherson, 1967; L'Ecuyer, 1988; Park and Miller, 1988.

19. Diaconis and Efron, 1983; Kolata, 1988; Peterson, 1991a; Efron and Tibshirani, 1993.

20. Browne, 1988; Oser, 1988; Peterson, 1988 (pp. 38–43, 214–217); Gleick, 1987a; Cipra, 1990.

21. Ford, 1983; Crutchfield et al., 1986; Gleick, 1987b.

9 Randomness as Uncertainty

1. Cicero, *De div.* 2.6.15, 2.9.24.

2. Hume, 1739 (pp. 125, 132, 135).

3. Mill, 1843 (p. 71). In subsequent editions, Mill backed down from this rash statement.

4. Venn, 1866, 3rd ed. (pp. xiii, 109).

5. Peirce, 1932 (p. 726).

6. Keynes, 1929 (p. 290).

7. The author of the entry on "Law of Error" was none other than F. Y. Edgeworth, who stated, "This *a priori* probability is sometimes grounded on our ignorance; according to another view, the procedure is justified by a rough general knowledge that over a tract of x for which P is sensible one value of x occurs about as often as another." Edgeworth, 1902 (p. 286).

8. Venn, 1866, 3rd ed. (p. 118).

9. Metropolis, Reitwiesner, and von Neumann, 1950.

10. On Eisenhart and Deming see Teichroew, 1965. See also Uhler, 1951; R. Greenwood, 1955; Wrench, 1960; Pathria, 1961, 1962 (p. 189); Stoneham, 1965. On Chudnovsky and Chudnovsky see MAA, 1989; Preston, 1992.

11. The digits of π play an interesting role in Carl Sagan's *Contact,* a science fiction novel about earth receiving messages from extraterrestrial life.

12. Von Mises, 1939 (pp. 20, 29, 30), named this random sequence a Kollectiv and defined a Kollectiv as "a long series of observations for which there are sufficient reasons to believe the hypothesis that the relative frequency of an attribute would tend to a fixed limit if it were indefinitely

continued." He called this fixed limit the *probability* of the attribute. Von Mises called the gambling systems "place selections" or "rules of place selection," because rules of prediction might incorporate the observation's position or place in the sequence. Patterned (nonrandom) sequences are often described by place or position; for instance, in the sequence 001001001001 . . . , every third digit is a 1, and all others are 0. For random sequences, rules of place selection would be impossible.

13. Keynes, 1929 (p. 290); Jeffreys, 1948, 1957. Excellent discussions of these objections and their ramifications can be found in von Mises, 1939; Popper, 1959; Loveland, 1966; Martin-Löf, 1969.

14. Von Mises, 1939 (p. 128); Reichenbach, 1949; Loveland, 1966; Martin-Löf, 1969.

15. Kolmogorov, 1963 (p. 369). The "law" or "algorithm" for prediction referred to by Kolmogorov corresponded to von Mises's place selection rules. See Sheynin, 1974.

16. Rather than the word "formula," Kolmogorov, 1965, used the word "program." "Program" meant "computer code," in binary digits, on an idealized computer. However, without loss of precision, one could substitute the words "mathematical formula" or "algorithm" for "program." Although Solomonoff, 1964, had earlier described similar measures to quantify the simplicity of hypotheses in models of induction, the definition of a random sequence in terms of informational complexity is credited independently to both Kolmogorov, 1965, 1968, and Gregory J. Chaitin, 1966, 1975.

17. Defining a random sequence according to its complexity removes predictably periodic sequences from the realm of randomness. However, this definition would also remove certain nonperiodic sequences which

can be described concisely such as the decimal expansion of certain irrational numbers like π.

18. According to Chaitin, 1975 (p. 48), "A series of numbers is random if the smallest algorithm capable of specifying it to a computer has about the same number of bits of information as the series itself." That is, the shortest formula that can specify a random sequence is almost as long as the sequence itself. This threshold of complexity below which a sequence would not be considered random remains arbitrary, and Chaitin has shown that even if the threshold is set rather high, "almost all" sequences of long finite length are random. Although it might seem relatively easy to find a random sequence if there are so many of them, Chaitin states that, under the strictest definition of informational complexity, it is impossible to do so.

19. Heidelberger, 1987.

20. Venn, 1866, 3rd ed. (p. 108).

21. See Wallis and Roberts, 1962; Hacking, 1965 (p. 123); Kirschenmann, 1972; Keene, 1957; Rescher, 1961 (p. 5).

22. Hacking, 1965 (p. 131).

23. Spencer Brown, 1957 (p. 149). See also Keene, 1957.

24. In the popular sense, "no particular order" means "disorderliness." Mathematically speaking, every sequence has "order" in that there is a first term, followed by a second term, followed by a third term, and so on. See Venn, 1866, 1st ed. (p. 6).

25. Fischbein and Gazit, 1984 (p. 17); Lopes, 1982; Kahneman and Tversky, 1982; Tversky and Kahneman, 1982, 1971.

26. Gage, 1943; Stoneham, 1965; Popper, 1959.

27. Humphreys, 1976; von Mises, 1939.

28. Kendall and Babington-Smith, 1938; Kendall, 1941; Keene, 1957.

29. Kendall and Babington-Smith, 1938, 1939a, 1939b.

30. Keene, 1957 (p. 157).

31. Kolmogorov, 1963.

32. Knuth, 1981.

33. Von Neumann, 1951 (p. 769).

10 Paradoxes in Probability

1. Polya, 1962; Kapadia and Borovcnik, 1991 (p. 2); Hawkins and Kapadia, 1984 (p. 359).

2. This problem is often referred to as "Mr. Smith and his son" or "Mr. Smith and his two sons." For an excellent discussion, see Bar-Hillel and Falk, 1982.

3. Also called the Prisoner's Paradox (not to be confused with the Prisoner's Dilemma). This version of the problem has been adapted from Ghahramani, 1996 (pp. 95–96).

4. Tierney, 1991 (p. A1).

5. See Shultz and Leonard, 1989; Kapadia and Borovcnik, 1991.

6. An adaptation of the example from Borovcnik, Bentz, and Kapadia, 1991 (pp. 66–67).

7. David, 1962; Sambursky, 1956.

8. Byrne, 1968 (p. 5).

9. See Kendall, 1956; David, 1962; Maistrov, 1974; Hacking, 1975; Daston, 1988; Patch, 1927 (p. 27).

10. Kendall, 1956 (p. 32).

Bibliography

Adrian, Robert. 1808. Research concerning the probabilities of the errors which happen in making observations, etc. *The Analyst, or Mathematical Museum* 1: 93–109.

American Academy of Pediatrics Committee on Infectious Diseases. 1994. Screening for tuberculosis in infants and children. *Pediatrics* 93: 131–134.

Ancient die goes to museum. 1931. *El Palacio* (Santa Fe) 31: 41.

Arbuthnot, John. 1714. *Of the laws of chance, or a method of Calculation of the hazards of game, plainly demonstrated, and applied to games at present most in use, which may be easily extended to the most intricate cases of chance imaginable.* London: Benj. Motte.

Ashton, John. 1893. *A history of English lotteries.* London: The Leaderhall Press. Rpt. Detroit: Singing Tree Press, 1969.

Avedon, Elliot M., and Brian Sutton-Smith, eds. 1971. *The study of games.* New York: John Wiley and Sons.

Bar-Hillel, Maya, and Ruma Falk. 1982. Some teasers concerning conditional probabilities. *Cognition* 11: 109–122.

Bernoulli, Daniel. 1777. The most probable choice between discrepant observations and the formation therefrom of the most likely induction. With an introductory note by M. G. Kendall. *Biometrika* 48: 1–18. Trans. C. G. Allen from *Acta Acad. Petrop.*, 1777, pp. 3–33. Rpt. in Pearson and Kendall, 1970, pp. 157–167.

Bhatta, C. Panduranga. 1985. *Dice play in Sanscrit literature.* Delhi: Amar Prakashan.

Bible. 1901. King James Version. Cleveland: World Publishing Company.

Bispham, J. W. 1920. An experimental determination of the distribution of the partial correlation coefficient in samples of thirty: Part I. Samples from an uncorrelated universe. *Proceedings of the Royal Society of London* 97: 218–224.

———— 1923. An experimental determination of the distribution of the partial correlation coefficient in samples of thirty: Samples from a highly correlated universe. *Metron* 2: 684–696.

Blake, William. 1825–1827. *Complete Writings.* Ed. Geoffrey Keynes. London: Nonesuch Press, 1957.

Bork, Alfred M. 1967. Randomness and the twentieth century. *Antioch Review* 27: 40–61.

Borovcnik, M., H.-J. Bentz, and R. Kapadia. 1991. A probabilistic perspective. In Kapadia and Borovcnik, 1991, pp. 27–71.

Boyer, Carl B. 1968. *A history of mathematics.* New York: John Wiley and Sons. Rpt. Princeton: Princeton University Press, 1985.

Browne, Malcolm W. 1988. Most ferocious math problem is tamed. *New York Times,* October 12, pp. A1, A17.

Buffon, Georges-Louis Leclerc. 1777. *Essai d'arthmetique morale.* Portions trans. John Lyon. In *From natural history to the history of nature: Readings from Buffon and his critics,* ed. and trans. John Lyon and

Phillip R. Sloan, pp. 53–73. Notre Dame: University of Notre Dame Press, 1981.

Buitenen, Johannes A. B. van, ed. and trans. 1975. *The Mahabharata.* Books II and III. Chicago: University of Chicago Press.

Burrill, Gail. 1990. Statistics and probability. *Mathematics Teacher* 83: 113–118.

Byrne, Edmund F. 1968. *Probability and opinion: A study in the medieval presuppositions of post-medieval theories of probability.* The Hague: Martinus Nijhoff.

Cardano, Girolamo. 1564. *Liber de ludo aleae.* [The book on games of chance.] Trans. Sydney H. Gould. Rpt. in Oystein Ore, *Cardano, the gambling scholar.* Princeton: Princeton University Press, 1953. (Page references are to Ore.)

Carlson, Karen J., Stephanie A. Eisenstat, and Terra Ziporyn. 1996. *The Harvard Guide to Women's Health.* Cambridge: Harvard University Press.

Carnarvon, George E., and Howard Carter. 1912. *Five years' explorations at Thebes.* London: Oxford University Press.

Cassells, Ward, Arno Schoenberger, and Thomas Grayboys. 1978. Interpretation by physicians of clinical laboratory results. *New England Journal of Medicine* 299: 999–1001.

Chaitin, Gregory J. 1966. On the length of programs for computing finite binary sequences. *Journal of the Association for Computing Machinery* 13: 547–569.

———— 1975. Randomness and mathematical proof. *Scientific American* 232: 47–52.

212   Chaucer, Geoffrey. 1949. *The Canterbury tales.* In *The portable Chaucer.*
Trans. and ed. Theodore Morrison. New York: Viking Press.

Church, Alonzo. 1940. On the concept of a random sequence. *American
Mathematical Society Bulletin* 46: 130–135.

Cicero, Marcus Tullius. 1920. *De divinatione.* Ed. and ann. Arthur S.
Pease. Urbana: University of Illinois Press.

——— 1928. *De divinatione.* Trans. William A. Falconer. London: William Heinemann Ltd.

Cioffari, Vincenzo. 1935. Fortune and fate from Democritus to St.
Thomas Aquinas. Ph.D. diss., Columbia University.

Cipra, Barry L. 1990. Computational complexity theorists tackle the
cheating computer conundrum. *SIAM (Society for Industrial and Applied Mathematicians) News* 23: 1, 18–20.

Clark, A. L. 1933. An experimental investigation of probability. *Canadian
Journal of Research* 9: 402–414.

——— 1937. Probability experimentally investigated. *Canadian Journal
of Research* 15: 149–153.

Cleary, Thomas, trans. 1986. *The Taoist I Ching.* Boston: Shambhala.

Coveyou, R. R. Quoted in Gardner, 1975, p. 169.

Coveyou, R. R., and R. D. MacPherson. 1967. Fourier analysis of uniform
random number generators. *Association for Computing Machinery
Journal* 14: 100–119.

Crutchfield, James P., J. Doyne Farmer, Norman H. Packard, and Robert
S. Shaw. 1986. Chaos. *Scientific American* 255: 46–57.

Culin, Stewart. 1896. Chess and playing cards. In *U.S. National Museum,
Annual Report.* Washington, DC: Smithsonian Institution.

———— 1903. American Indian games. *American Anthropologist*, n.s. 5: 58–64. Rpt. in Avedon and Sutton-Smith, 1971, pp. 103–107.

———— 1907. *Games of the North American Indians*. Washington, DC: Government Printing Office.

Darbishire, A. D. 1907. Some tables for illustrating statistical correlation. *Manchester Literary and Philosophical Society, Memoirs and Proceedings* 51: 1–20.

Daston, Lorraine. 1988. *Classical probability during the Enlightenment*. Princeton: Princeton University Press.

David, Florence N. 1962. *Games, gods and gambling: Origins and history of probability and statistical ideas from the earliest times to the Newtonian era*. New York: Hafner Publishing Company.

Davidson, Henry A. 1949. *A short history of chess*. New York: Greenberg Publisher.

DeForest, Erastus L. 1876. *Interpolation and adjustment of series*. New Haven: Tuttle, Morehouse & Taylor. Rpt. in Stigler, 1980, vol. 2.

De Moivre, Abraham. 1756. *The doctrine of chances: Or, a method of calculating the probabilities of events of play*. 3d ed. London: A. Millar. Rpt. New York: Chelsea Publishing Company, 1967.

De Morgan, Augustus. 1912. *A budget of paradoxes*. 2nd ed. Ed. David Eugene Smith. Freeport, N.Y.: Books for Libraries Press, 1969.

De Vreese, K. 1948. The game of dice in ancient India. *Orientalia Neerlandica*. 25th anniversary volume: 349–362.

Diaconis, Persi. 1989. "The search for randomness." Talk, March 29. Teachers College, Columbia University.

Diaconis, Persi, and Bradley Efron. 1983. Computer-intensive methods in statistics. *Scientific American* 248: 116–130.

214 Diaconis, Persi, and Frederick Mosteller. 1989. Methods for studying coincidences. *Journal of the American Statistical Association* 84: 853–861.

Edgeworth, Francis Y. 1885a. Methods of statistics. *Journal of the Royal Statistical Society* Jubilee Volume: 181–217.

———— 1885b. On methods of ascertaining variations in the rate of births, deaths, and marriages. *Journal of the Statistical Society* 48: 628–649.

———— 1902. Error, Law of. In *Encyclopaedia Britannica.* 10th ed. London: Adam and Charles Black.

Efron, Bradley, and Robert J. Tibshirani. 1993. *An introduction to the bootstrap.* Monograph on Statistics and Applied Probability 57. New York: Chapman and Hall.

Epstein, Rabbi I., ed. 1935. *The Babylonian Talmud.* Trans. Jacob Shachter. London: The Soncino Press.

Erasmus, C. J. 1971. Patolli, pachisi, and the limitations of possibilities. In Avedon and Sutton-Smith, 1971, pp. 109–129.

Evans, Arthur J. 1964. *The palace of Minos at Knossos.* New York: Biblo and Tannen.

Ewin, C. L'Estrange. 1972. *Lotteries and sweepstakes.* New York: Benjamin Blom.

Ezell, John S. 1960. *Fortune's merry wheel: The lottery in America.* Cambridge: Harvard University Press.

Falkener, Edward. 1892. *Games ancient and oriental and how to play them.* London: Longmans, Green and Company. Rpt. New York: Dover Publications, 1961.

Fienberg, Stephen E. 1971. Randomization and social affairs: The 1970 draft lottery. *Science* 171: 255–261.

Fischbein, Efraim, and A. Gazit. 1984. Does the teaching of probability improve probabilistic intuitions? *Educational Studies in Mathematics* 15: 1–24.

Fisher, Ronald A. 1926. On the random sequence. *Quarterly Journal of the Royal Meteorological Society* 52: 250.

Fisher-Schreiber, Ingrid, Franz-Karl Ehrhard, Kurt Friedrichs, and Michael S. Diener. 1989. *The encyclopedia of Eastern philosophy and religion: Buddhism, Hinduism, Taoism, and Zen.* Boston: Shambhala. S.v. "Hung-fan," "I Ching," and "Kali."

Ford, Joseph. 1983. How random is a coin toss? *Physics Today* 36: 40–47.

Gadd, C. J. 1934. An Egyptian game in Assyria. *Iraq* 1: 45–50.

Gage, Robert. 1943. Contents of Tippett's "Random sampling numbers." *Journal of the American Statistical Association* 38: 223–227.

Galilei, Galileo. 1623. *Sopra le scoperte dei dadi.* [Thoughts about dice games.] In vol. 8, *Opere di Galileo Galilei.* Trans. E. H. Thorne. Florence: Edizione Nationale, 1898. Rpt. in David, 1962, pp. 192–195.

——— 1632. *Dialogue of the two chief world systems—Ptolemaic and Copernican.* Third day discussion. Trans. S. Stillman Drake. With a Foreword by Albert Einstein. Berkeley: University of California Press, 1953.

Galton, Francis. 1877. Typical laws of heredity. *Nature* 15: 492–495, 512–514, 532–533.

——— 1890a. Kinship and correlation. *North American Review* 150: 419–431. Rpt. in Stigler, 1989, pp. 81–86.

——— 1890b. Dice for statistical experiments. *Nature* 42: 13–14. Rpt. in Stigler, 1991, pp. 94–96.

216

Gardner, Martin. 1975. Random numbers. In *Mathematical carnival*. New York: Alfred A. Knopf. Rpt. New York: Vintage Books, 1977.

Gauss, Karl Friedrich. 1809. *Theoria motus corporum coelestium.* [Theory of the motion of the heavenly bodies.] Trans. Charles H. Davis. Boston: Little, Brown and Company, 1857.

Ghahramani, Saeed. 1996. *Fundamentals of probability.* Upper Saddle River, NJ: Prentice Hall.

Gleick, James. 1987a. A new approach to protecting secrets is discovered. *New York Times*, February 17, pp. C1, C3.

―――― 1987b. *Chaos.* New York: Viking.

Goldstein, Kenneth S. 1971. Strategy in counting-out: An ethnographic folklore field study. In Avedon and Sutton-Smith, 1971, pp. 167–178.

Gosset, William S. See Student.

Graves, Robert. 1934. *I, Claudius.* New York: Harrison Smith and Robert Haas.

Greenwood, Robert E. 1955. Coupon collector's test for random digits. *Mathematical Tables and Other Aids to Computation* 9: 1–5.

Gridgeman, N. T. 1960. Geometric probability and the number π. *Scripta Mathematica* 25: 183–195.

Grierson, George A. 1904. Guessing the number of the *vibhitaka* nuts. *Journal of the Royal Asiatic Society of Great Britain and Ireland*: 355–357.

Hacking, Ian. 1965. *Logic of scientific inference.* Cambridge: Cambridge University Press.

―――― 1975. *The emergence of probability: A philosophical study of early*

ideas about probability, induction and statistical inference. New York: **217**
Oxford University Press.

Hall, A. 1873. On an experimental determination of π. *Messenger of Mathematics* 2: 113–114.

Hasofer, A. M. 1967. Random mechanisms in Talmudic literature. *Biometrika* 54: 316–321. Rpt. in Pearson and Kendall, 1970, pp. 39–44.

Hastings, James, ed. 1912. *Encyclopedia of religion and ethics.* New York: Charles Scribner's Sons. S.v. "Chance," "City, city-gods," "Divination," "Fortune," "Gambling," and "Games."

Hawkins, Anne S., and Ramesh Kapadia. 1984. Children's conceptions of probability: A psychological and pedagogical review. *Educational Studies in Mathematics* 15: 349–377.

Heidelberger, Michael. 1987. Fechner's indeterminism: From freedom to laws of chance. In Krüger, Daston, and Heidelberger, 1987, pp. 117–156.

Held, G. J. 1935. Gambling. Chap. 5. in *The Mahabharata: An ethnological study.* London: Kegan Paul, Trench, Trubner and Co.

Herodotus. 1859. *The history of Herodotus.* Trans. and ed. George Rawlinson. New York: D. Appleton & Co.

Hillerman, Tony. 1990. *Coyote waits.* New York: Harper & Row.

Hobbes, Thomas. 1841. *The questions concerning liberty, necessity, and chance.* Vol. 5 in *The English works of Thomas Hobbes.* Ed. Sir William Molesworth. London: John Bohn.

Homer. 1951. *The Iliad.* Trans. Richmond Lattimore. Chicago: University of Chicago Press.

218 Horton, H. Burke. 1948. A method for obtaining random numbers. *Annals of Mathematical Statistics* 19: 81–85.

Horton, H. Burke, and R. Tynes Smith, III. 1949. A direct method for producing random digits in any number system. *Annals of Mathematical Statistics* 20: 82–90.

Hume, David. 1739. *A treatise of human nature: Being an attempt to introduce the experimental method of reasoning into moral subjects,* vol. 1, part 3: *Of the understanding of knowledge and probability.* Ed. L. A. Selby-Bigge. London: Oxford University Press, 1965.

Humphreys, Paul William. 1976. Inquiries in the philosophy of probability: Randomness and independence. Ph.D. diss., Stanford University.

Jackson, Shirley. 1948. The lottery. In *Literature and the writing process.* 3rd ed. New York: Macmillan Company, 1993.

Jeffreys, Harold. 1948. *The theory of probability.* 2nd ed. London: Oxford University Press.

———— 1957. *Scientific inference.* 2nd ed. Cambridge: Cambridge University Press.

Kahneman, Daniel, Paul Slovic, and Amos Tversky, eds. 1982. *Judgement under uncertainty: Heuristics and biases.* New York: Cambridge University Press.

Kahneman, Daniel, and Amos Tversky. 1982. Subjective probability: A judgment of representativeness. In Kahneman, Slovic, and Tversky, 1982, pp. 32–47.

Kapadia, Ramesh, and Manfred Borovcnik, eds. 1991. *Chance encounters: Probability in education.* The Netherlands: Kluwar Academic Publ.

Keene, G. B. 1957. Randomness II. *The Aristotelian Society Symposium*

Proceeding, 12–14 July 1957; Supplementary 31: 151–160. London: **219**
Harrison and Sons, Ltd.

Kendall, Maurice G. 1941. A theory of randomness. *Biometrika* 32: 1–15.

———— 1956. The beginnings of a probability calculus. *Biometrika* 43: 1–14. Rpt. in Pearson and Kendall, 1970, pp. 19–34.

———— 1961. Daniel Bernoulli on maximum likelihood. Introductory note to Bernoulli, 1777. Rpt. in Pearson and Kendall, 1970, pp. 155–156.

Kendall, Maurice G., and B. Babington-Smith. 1938. Randomness and random sampling numbers. *Journal of the Royal Statistical Society,* ser. A, 101: 147–166.

———— 1939a. Second paper of random sampling numbers. *Journal of the Royal Statistical Society,* ser. B, 6: 51–61.

———— 1939b. Tables of random sampling numbers. *Tracts for Computers,* no. 24. Rpt. London: Cambridge University Press, 1951.

Kendall, Maurice G., and R. L. Plackett, eds. 1977. *Studies in the history of statistics and probability,* vol. 2. London: Charles Griffin & Co.

Keynes, John M. 1929. The meanings of objective chance, and of randomness. Chap. 24 in *A treatise on probability.* London: Macmillan & Co.

Kirschenmann, Peter. 1972. Concepts of randomness. *Journal of Philosophical Logic* 1: 395–413.

Knuth, Donald E. 1981. Random numbers. Chap. 3 in *Seminumerical algorithms,* vol. 2 of *The art of computer programming.* 2nd ed. Reading, MA: Addison Wesley.

Kolata, Gina. 1988. Theorist applies power to uncertainty in statistics. *New York Times,* November 8, pp. C1, C6.

——— 1990. 1-in-a-trillion coincidence, you say? Not really, experts find. *New York Times,* February 27, pp. C1–C2.

Kolmogorov, Andrei. 1963. On tables of random numbers. *Sankhya,* ser. A 25: 369–376.

——— 1965. Three approaches for defining the concept of information quantity. *Problems of Information Transmission* 1: 4–7.

——— 1968. Logical basis for information theory and probability theory. *IEEE Transactions on Information Theory* IT-14: 662–664.

Krüger, Lorenz, Lorraine Daston, and Michael Heidelberger, eds. 1987. *The probabilistic revolution,* vol. 1: *Ideas in history.* Cambridge: MIT Press.

Kruskal, William H., and Frederick Mosteller. 1980. Representative sampling, IV: The history of the concept in statistics, 1895–1939. *International Statistical Review* 48: 169–195.

Langdon, Steven H. 1930. The Semitic goddess of fate, Fortuna-Tyche. *Journal of the Royal Asiatic Society of Great Britain and Ireland:* 21–29.

——— 1931. *The mythology of all races.* Vol. 5, *Semitic.* Rpt. New York: Cooper Square Publishing, Inc., 1964.

Laplace, Pierre-Simon de. 1886. *Oeuvres complètes de Laplace,* vol. 7, book II, chaps. 4 and 5. Paris: Gauthier-Villars.

L'Ecuyer, Pierre. 1988. Efficient and portable combined random number generators. *Communications of the Association of Computing Machinery* 31: 742–749, 774.

Le Guin, Ursula K. 1969. *The left hand of darkness.* New York: Harper & Row.

Lehmer, D. H. 1951. Mathematical methods in large-scale computing units. In *The Annals of the Computation Laboratory of Harvard Uni-*

versity 26: Proceedings of a second symposium on large-scale digital calculating machinery, 13–16 September 1949.

Lichtenstein, Murray, and Louis I. Rabinowitz. 1972. *Encyclopaedia Judaica.* New York: Macmillan. S.v. "Gambling," "Games," and "Lots."

Lopes, Lola L. 1982. Doing the impossible: A note on induction and the experience of randomness. *Journal of Experimental Psychology* 8: 626–636.

Loveland, Donald. 1966. A new interpretation of the von Mises concept of random sequences. *Zeitschrift für Mathematische Logik und Grundlagen der Mathematik* 12: 179–194.

Lucretius. 1937. *De rerum natura,* vol. 1, book 2. 3rd ed. Trans. W. H. D. Rouse. London: William Heinemann.

MAA. 1989. *See* Mathematical Association of America.

Mackay, Ernest J. 1976. *Further excavations at Mohenjo-Daro.* 2 vols. New Delhi: Indological Book Corp.

Mahabharata. See Buitenen, 1975.

Maistrov, Leonid E. 1974. *Probability theory: A historical sketch.* Trans. Samuel Kotz. New York: Academic Press.

Marsaglia, George, and Arif Zaman. 1991. A new class of random number generators. *Annals of Applied Probability* 1 (3): 462–480.

Martin-Löf, Per. 1969. The literature on von Mises Kollectivs revisited. *Theoria* 35: 12–37.

Massey, William A. 1996. Correspondence with J. Laurie Snell, ed. *Chance News,* January 9, 1996.

Mathematical Association of America (MAA). 1989. At the limits of calculation: Pi to a billion digits and more. *Focus* 9: 1, 3–4.

222   Metropolis, N. C., G. Reitwiesner, and J. von Neumann. 1950. Statistical treatment of values of first 2000 decimal digits of e and π calculated on the ENIAC. *Mathematical Tables and Other Aids to Computation* 4: 109–111.

Meyer, Herbert, ed. 1956. *Symposium on Monte Carlo methods.* University of Florida, Statistical Lab. New York: John Wiley and Sons, Inc.

Mill, John Stuart. 1843. *A system of logic, ratiocinative and inductive, being a connected view of the principles of evidence, and the methods of evidence, and the methods of scientific investigation,* vol. 2. London: John W. Parker.

Mises, Richard von. 1939. *Probability, statistics and truth.* 2nd ed. Trans. J. Neyman, D. Sholl, and E. Rabinowitsch. New York: Macmillan.

Neumann, J. von. 1951. Various techniques used in connection with random digits. *Journal Res. Nat. Bus. Stand. Appl. Math. Series* 3: 36–38. Rpt. in *John von Neumann, collected works,* vol. 5, pp. 768–770, ed. A. H. Taub. New York: Macmillan, 1963.

News and views. 1929. *Nature* 123: 540.

Nisbett, Richard E., Eugene Borgida, Rick Crandall, and Harvey Reed. 1982. Popular induction: Information is not necessarily informative. In Kahneman, Slovic, and Tversky, 1982, pp. 101–116.

OED. 1989. *See Oxford English dictionary.*

Oppenheim, A. Leo. 1970. *Ancient Mesopotamia: Portrait of a dead civilization.* Chicago: University of Chicago Press.

Oppenheim, A. Leo, Ignace J. Gelb, and Benno Landsberger, eds. 1960. *The Assyrian dictionary.* Chicago: Oriental Institute. S.v. "Isqu."

Ore, Oystein. 1953. *Cardano, the gambling scholar.* Princeton: Princeton University Press.

——— 1960. Pascal and the invention of probability theory. *American* **223**
Mathematical Monthly 67: 409–419.

Oser, Hans J. 1988. Team from Sandia Laboratories puts the factoring problem on a hypercube. *SIAM (Society for Industrial and Applied Mathematicians) News* (September): 1, 15.

Oxford English dictionary (OED). 1989. 2nd ed. Ed. J. A. Simpson and E. S. C. Weiner. Oxford: Clarendon Press. S.v. "Chance," "Cleromancy," "Cube," "Hazard," "Lot," "Lottery," "Random," "Rhapsodomancy," "Sortes," "Sortilege," and "Stochasic."

Park, Stephen K., and Keith W. Miller. 1988. Random number generators: Good ones are hard to find. *Computing Practices* 31: 1192–1201.

Parke, H. W. 1988. *Sibyls and Sibylline prophecy in classical antiquity.* Ed. B. C. McGing. London and New York: Routledge.

Pasteur, Louis. 1854. Quoted in Vallery-Radot, 1927.

Patch, Howard R. 1927. *The goddess Fortuna in mediaeval literature.* Cambridge: Harvard University Press.

Pathria, R. K. 1961. A statistical analysis of the first 2,500 decimal places of e and $1/e$. *Proceedings of the National Institute of Sciences of India,* Part A 27: 270–282.

——— 1962. A statistical study of randomness among the first 10,000 digits of π. *Mathematics of Computation* 16: 188–197.

Pearson, Egon S. 1939. Foreword to *Tables of random sampling numbers: Tracts for computers,* no. 24. In Kendall and Babington-Smith, 1939b.

——— 1965. Some incidents in the early history of biometry and statistics, 1890–1894. *Biometrika* 52: 3–18. Rpt. in Pearson and Kendall, 1970, pp. 323–338.

——— 1967. Some reflections on continuity in the development of mathematical statistics, 1885–1920. *Biometrika* 54: 341–355. Rpt. in Pearson and Kendall, 1970, pp. 339–354.

Pearson, Egon S., and Maurice G. Kendall, eds. 1970. *Studies in the history of statistics and probability.* London: Charles Griffin & Co.

Pearson, Karl. 1895. Contributions to the mathematical theory of evolution, II. Skew variation in homogeneous material. *Philosophical Transactions of the Royal Society of London* (A) 186: 343–414.

———— 1900. On the criterion that a given system of deviations from the probable in the case of a correlated system of variables is such that it can be reasonably supposed to have arisen from random sampling. *London, Edinburgh and Dublin Philosophical Magazine and Journal of Science* 50: 157–175.

———— 1924. Historical note on the origin of the normal curve of errors. *Biometrika* 16: 402–404.

Pease, Arthur S., ed. 1920. *Cicero's De divinatione.* Urbana: University of Illinois.

Peatman, John Gray, and Roy Shafer. 1942. A table of random numbers from selective service numbers. *Journal of Psychology* 14: 295–305.

Peirce, Charles Sanders. 1932. *Collected papers of Charles Sanders Peirce.* Ed. Charles Hartshorne and Paul Weiss. Vol. 2, *Elements of logic.* Cambridge: Harvard University Press.

Peirce, Charles Sanders, and Joseph Jastrow. 1884. On small differences of sensation. *Memoirs of the National Academy of Sciences for 1884* 3: 75–83. Rpt. in Stigler, 1980, vol. 2.

Peterson, Ivars. 1988. *The Mathematical Tourist.* New York: W. H. Freeman.

———— 1991a. Pick a sample. *Science News* 140: 56–58.

———— 1991b. Numbers at random. *Science News* 140: 300–301.

Petrie, W. M. Flinders, and Guy Brunton. 1924. *Sedment I.* London: British School of Archaeology in Egypt.

Petrie, W. M. Flinders, and James E. Quibell. 1896. *Nagada and Ballas. 1895.* British School of Archaeology in Egypt. London: Bernard Quaritch.

Piaget, Jean, and Barbel Inhelder. 1975. *The origin of the idea of chance in children.* Trans. Lowell Leake, Jr., Paul Burrell, and Harold D. Fishbein. New York: W. W. Norton.

Polya, George. 1962. *Mathematical discovery.* New York: John Wiley and Sons.

Popper, Karl R. 1959. *The logic of scientific discovery.* London: Hutchinson & Co.

Porter, Theodore M. 1986. *The rise of statistical thinking, 1820–1900.* Princeton: Princeton University Press.

Preston, Richard. 1992. Profiles: The mountains of Pi (David and Gregory Chudnovsky). *New Yorker,* March 2: 36–67.

Quibell, James E. 1913. *Egypt: Excavations at Saqqara; Tombs of Hesy. Service des antiquites, 1911–12.* Cairo: Imprimerie de l'Institut Francais.

Rabinovitch, Nachum L. 1973. *Probability and statistical inference in ancient and medieval Jewish literature.* Toronto: University of Toronto Press.

RAND Corporation. 1955. *A million random digits with 100,000 normal deviates.* Glencoe, Ill.: Free Press.

Reichenbach, Hans. 1949. *The theory of probability, and inquiry into the logical and mathematical foundations of the calculus of probability.* 2nd

226

ed. Trans. Ernest H. Hutten and Maria Reichenbach. Berkeley: University of California Press.

Remington, J. S., and G. R. Hollingworth. 1995. New tuberculosis epidemic: Controversies in screening and preventative therapy. *Canadian Family Physician* 41: 1014–1023.

Report on second NBSINA Symposium. 1948. *Mathematical Tables and Other Aids to Computation* 3: 546.

Rescher, Nicholas. 1961. The concept of randomness. *Theoria* 27: 1–11.

Rhodes, Richard. 1995. *Dark sun: The making of the hydrogen bomb.* New York: Simon and Schuster.

Ronan, Colin A. 1967. *Their majesties' astronomers.* Great Britain: The Bodley Head Ltd. Rpt. as *Astronomers royal.* Garden City, NY: Doubleday, 1969.

Sam:bursky, S. 1956. On the possible and probable in ancient Greece. *Osiris* 12: 35–48. Rpt. in Kendall and Plackett, 1977, pp. 1–14.

——— 1959a. *Physics of the Stoics.* New York: Macmillan Co.

——— 1959b. *The physical world of the Greeks.* Trans. Merton Dagut. London: Routledge and Kegan Paul. Rpt. 1963.

Shafer, Glenn. 1978. Non-additive probabilities in the work of Bernoulli and Lambert. *Archive for History of Exact Sciences* 19: 309–370.

Shchutskii, Iulian K. 1979. *Researches on the I Ching.* Trans. William L. MacDonald, Tsuyoshi Hasegawa, and Hellmut Wilhelm. Princeton: Princeton University Press.

Sheynin, O. B. 1968. On the early history of the law of large numbers. *Biometrika* 55: 459–467. Rpt. in Pearson and Kendall, 1970, pp. 231–239.

———— 1971. Newton and the classical theory of probability. *Archive for History of Exact Sciences* 7: 217–243.

———— 1974. On the prehistory of the theory of probability. *Archive for History of Exact Sciences* 12: 97–141.

Shultz, Harris S., and Bill Leonard. 1989. Probability and Intuition. *Mathematics Teacher* 82: 52–53.

Simpson, Thomas. 1756. A letter to the Right Honourable George Earl of Macclesfield, President of the Royal Society, on the advantage of taking the mean of a number of observations, in practical astronomy. *Philosophical Transactions of the Royal Society of London* 49: 82–93.

Smith, William, William Wayte, and G. E. Marindin, eds. 1901. *A dictionary of Greek and Roman antiquities*. 3rd ed. London: John Murray. S.v. "Alea," "Divinatio," "Duodecim scrita," "Fritillus," "Latrunculi," "Micare digitis," "Oraculum," "Par impar ludere," "Sibyllini libri," "Situla or sitella," "Sortes," "Talus," and "Tessera."

Solomonoff, Ray J. 1964. A formal theory of inductive inference. Part I. *Information and Control* 7: 1–22.

Spencer Brown, G. 1957. *Randomness I.* The Aristotelian Society Symposium Proceeding, 12–14 July 1957; Supplementary 31: 145–150. London: Harrison and Sons.

Stigler, Stephen M. 1978. Mathematical statistics in the early states. *Annals of Statistics* 6: 239–265.

———— 1980. *American contributions to mathematical statistics.* 2 vols. New York: Arno Press.

———— 1986. *The history of statistics: The measurment of uncertainty before 1900.* Cambridge: Harvard University Press.

228

—— 1989. Francis Galton's account of the invention of correlation. *Statistical Science* 4: 73–86.

—— 1991. Stochastic simulation in the nineteenth century. *Statistical Science* 6: 89–97.

Stoneham, R. G. 1965. A study of 60,000 digits of the transcendental e. *American Mathematical Monthly* 72: 483–500.

Student. [William S. Gosset.] 1908a. The probable error of a mean. *Biometrika* 6: 1–25.

—— 1908b. Probable error of a correlation coefficient. *Biometrika* 6: 302–310.

Suetonius. 1914. *De vita Caesarum.* [The lives of the Caesars.] Trans. J. C. Rolfe. London: William Heinemann. Rpt. 1924.

Sullivan, George. 1972. *By chance a winner: The history of lotteries.* New York: Dodd, Mead and Company.

Tacitus, Cornelius. 1948. *Germania.* Trans. H. Mattingly, rev. S. A. Hanford. Middlesex: Penguin Books, 1970.

Teichroew, Daniel. 1965. A history of distribution sampling prior to the era of the computer and its relevance to simulation. *Journal of the American Statistical Association* 60: 27–49.

Tierney, John. 1991. Behind Monty Hall's doors: Puzzle, debate and answer? *New York Times,* July 21, pp. A1, A20.

Tippett, Leonard H. C. 1925. On the extreme individuals and the range of samples taken from a normal population. *Biometrika* 17: 364–387.

—— 1927. *Random sampling numbers: Tracts for computers,* No. 15. With a Foreword by Karl Pearson. London: Cambridge University Press, 1950.

Todhunter, Isaac. 1865. *A history of the mathematical theory of probability from the time of Pascal to that of Laplace.* London: Macmillan. Rpt. New York: G. E. Stechert and Co., 1931.

Tversky, Amos, and Maya Bar-Hillel. 1983. Risk: The long and the short. *Journal of Experimental Psychology: Learning, Memory, and Cognition* 9: 713–717.

Tversky, Amos, and Daniel Kahneman. 1971. The belief in the law of small numbers. *Psychological Bulletin* 76: 105–110.

—— 1974. Judgement under uncertainty: Heuristics and biases. *Science* 185: 1124–1131.

—— 1982. Evidential impact of base rates. In Kahneman, Slovic, and Tversky, 1982, pp. 153–162.

Tylor, Edward B. 1873. *Primitive culture,* vol. 1. 2nd ed. London: John Murray. Rpt. as *The origins of culture.* New York: Harper & Row, 1958.

—— 1879. Lecture: The history of games. *Proceedings of the Royal Institution* 9: 125–139. Rpt. in Avedon and Sutton-Smith, 1971, pp. 63–76.

—— 1896. On American lot-games as evidence of Asiatic intercourse before the time of Columbus. *International Archives for Ethnographie* Supplement 9: 56–66. Rpt. in Avedon and Sutton-Smith, 1971, pp. 77–93.

Uhler, H. S. 1951. Many-figure approximations to $\sqrt{2}$, and distribution of digits in $\sqrt{2}$ and $1/\sqrt{2}$. *Proceedings of the National Academy of Sciences of the United States of America* 37: 63–67.

Vallery-Radot, René. 1927. *The Life of Pasteur.* Trans. R. L. Devonshire. Garden City, NY: Garden City Publishing Co.

Venn, John. 1866. *The logic of chance: An essay on the foundations and*

230

province of the theory of probability, with especial reference to its application to moral and social science. London: Macmillan. 2nd ed., 1876; 3rd ed., 1888; 4th ed., 1962. Rpt. Chelsea Publishing Co.

Vickery, C. W. 1939. On drawing a random sample from a set of punched cards. *Journal of the Royal Statistical Society* Supplement 6: 62–66.

Virgil. 1952. *The Aeneid.* Trans. C. Day Lewis. Garden City, NY: Doubleday.

Waddell, L. Austine. 1939. *Buddhism of Tibet or Lamaism.* 2nd ed. Cambridge: W. Heffer and Sons, Ltd. Rpt., 1971.

Walker, Benjamin. 1968. *The Hindu world: An encyclopedic survey of Hinduism.* New York: Frederick A. Praeger. S.v. "Gambling" and "Mahabharata."

Walker, Helen M. 1929. *Studies in the history of statistical method, with special reference to certain educational problems.* Baltimore: Williams and Wilkins.

Wallis, W. Allen, and Henry V. Roberts. 1962. Randomness. Chap. 6 in *The nature of statistics.* 2nd ed. New York: The Free Press.

Weaver, Warren. 1963. *Lady Luck: The theory of probability.* New York: Dover Publishing, Inc.

Weldon, Walter F. R. 1906. Inheritance in animals and plants. In *Lectures on the method of science,* ed. T. B. Strong, pp. 81–109. London: Oxford University Press.

Wilhelm, Richard. Trans. 1950. *I Ching.* Trans. Cary F. Baynes. New York: Pantheon Books.

Winternitz, Moriz. 1981. *Geschichte der Indischen literatur.* [A history of Indian literature.] Trans. V. Srinivasa Sarma. Delhi: Motilal Banarsidass.

Woolley, C. Leonard. n.d. *Excavations at Ur.* New York: Thomas Y. Crowell Co.

———— 1928. Excavations at Ur, 1927–28. *Antiquaries Journal* 8: 415–448.

Wrench, J. W., Jr. 1960. The evolution of extended decimal approximations to π. *Mathematics Teacher* 53: 644–650.

Index